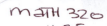

Ordinary Differential Equations
using MATLAB®

THIRD EDITION

John C. Polking
David Arnold

PEARSON

Prentice
Hall

Upper Saddle River, NJ 07458

Editor-in-Chief: Sally Yagan
Acquisitions Editor: George Lobell
Supplement Editor: Jennifer Brady
Assistant Managing Editor: John Matthews
Production Editor: Allyson Kloss
Supplement Cover Manager: Paul Gourhan
Supplement Cover Designer: Joanne Alexandris
Manufacturing Buyer: Ilene Kahn

© 2004 Pearson Education, Inc.
Pearson Prentice Hall
Pearson Education, Inc.
Upper Saddle River, NJ 07458

Printed in the United States of America

10 9 8 7 6

ISBN 0-13-145679-2

Pearson Education Ltd., *London*
Pearson Education Australia Pty. Ltd., *Sydney*
Pearson Education Singapore, Pte. Ltd.
Pearson Education North Asia Ltd., *Hong Kong*
Pearson Education Canada, Inc., *Toronto*
Pearson Educación de Mexico, S.A. de C.V.
Pearson Education—Japan, *Tokyo*
Pearson Education Malaysia, Pte. Ltd.

Contents

Preface

This, the third edition of *Ordinary Differential Equations, Using MATLAB*, *
responds to the release of MATLAB 6 by the Mathworks. It describes new editions
of `dfield` and `pplane`, now called `dfield6` and `pplane6`. These programs have
been rewritten to take advantage of new features in MATLAB. Some new features
have been added that users have requested. It also describes the new program,
`odesolve`, which wraps a graphical user interface around MATLAB's suite of ODE
solvers.

All of the MATLAB commands in this edition have been thoroughly tested
with version 6.0 of MATLAB and version 3.0.1 of the Symbolic Toolbox, as well
as with later releases. If you are currently using MATLAB 5, then you will either
need to upgrade your software or consider using the second edition of *Ordinary
Differential Equations, Using MATLAB*. If your version of the Symbolic Toolbox is
a release prior to version 3.0.1, you can still use this manual, but the responses to
the commands might differ. You should consider upgrading to the latest version.

We still feel that there is no elementary college level mathematics course for
which computer graphics is more useful to the student than the ordinary differential
equations course. This manual reflects our ongoing efforts to institute a significant
computer component in the ODE courses taught at Rice University and College of
the Redwoods.

The manual is designed to be used by the student while working at the computer.
The student should follow the manual, executing the commands as they occur in the
text. Pains have been taken to make it as easy as possible for the student to work
through the manual with a minimum of assistance. Of course, some assistance will
inevitably be required.

This manual is meant to be a supplementary text for students in an ODE course.
It is not a textbook, nor was it written as a companion for any particular textbook.
In fact, this manual can accompany most ODE textbooks (we'd like to think all),
and we leave the choice of text to the instructor. Although the manual is basically a

---

* MATLAB is a registered trademark of
   The MathWorks, Inc.
   For MATLAB product information, please contact
   The MathWorks, Inc.
   24 Prime Park Way
   Natick, MA 01760-1500
   Phone: 508-647-7000
   Fax: 508-647-7101
   Email: info@mathworks.com
   Web: www.mathworks.com

book on how to use and solve ODEs with MATLAB, there are theorems stated, and there are theoretical results quoted in the manual. However, there are no proofs.

With a plethora of excellent ODE software packages to choose from, we have often been asked why we chose MATLAB for our students. The answer is that we wanted our students to spend their time learning about ODEs and not about software. That meant that the software used should be as easy as possible to use. Another consideration, possibly in conflict with the first, was that if our students were going to spend time learning software, then it would be desirable if that knowledge were useful in other venues after the course was over. After looking at several mathematical software packages, we decided that MATLAB was by far the easiest to use. In addition MATLAB's suite of ODE solvers provide extraordinary power to solve systems of ODEs numerically. So we chose MATLAB, and we have been pleased with the results. By all indications, our students have been pleased as well.

Although the emphasis in the manual is still on ODEs, we do take more pains to explain the many features of MATLAB in this edition than we did in the first two. Our reason for doing so is to enable students to solve more difficult problems and prepare more sophisticated graphics as part of ODE projects. Thus, the reader will learn a great deal about MATLAB while working through this manual. The only programming in the manual is an occasional use of a simple for loop.

The first chapter is extremely elementary. It is meant for the user with almost no experience with computers. In a departure from previous editions, Chapter 2 is entirely devoted to the use of MATLAB to plot and format the graphs of functions, especially of the solutions of ODEs that students encounter early in their ODE course. For this it is necessary to understand how MATLAB works with vectors and how to write script M-files of minimal complexity. We find that students learn this material easily.

We believe that the early introduction of computer drawn direction fields and solution trajectories is an essential ingredient in a modern course on differential equations. The MATLAB function dfield6, which displays the direction field of a single, user-provided ODE of first order, is very easy to use, and does not require any knowledge of how MATLAB works beyond knowing how to start it. We like to get our students acquainted with dfield6 as early as the first day of instruction. Its use is described in Chapter 3.

The use of MATLAB function M-files greatly expands the ability of the user to do computational exercises and prepare graphics. They are also needed for the solutions of ODES described in Chapters 5 and 8. All of Chapter 4 is devoted to the use of function M-files.

Chapter 5 uses MATLAB to explore numerical methods of solving ODEs. Using MATLAB, it is quite easy to explore the errors actually made when the standard one-step solution methods due to Euler and Runge-Kutta are used.

In Chapter 6 we discuss advanced uses of the dfield6 routine introduced in Chapter 3. Engineering students will find this chapter particularly useful, because the chapter highlights and explains forcing functions typically encountered in engineering courses (square waves, pulse trains, etc.).

Systems of ODEs are being introduced to students much earlier in the modern ODE course. Chapter 7 introduces `pplane6`, a routine designed to explore planar, autonomous systems of ODEs. We put off advanced discussion of `pplane6` until Chapter 13.

Chapter 8 is one of the most important chapters in the manual. It is here that MATLAB's suite of ODE solvers is described, and where the user will learn how to solve essentially any system of ODEs numerically. Learning this skill should be a goal of a modern ODE course. Students of science and engineering should be able to use this tool in later courses and in job experiences.

In Chapter 9 we describe `odesolve`, our new tool for solving systems of ODEs. The tool wraps a graphical user interface around the MATLAB suite of solvers, thereby facilitating the use of the suite to solve essentially any system of ODEs. This is the first edition of `odesolve`, and it is a little crude. The authors are interested in how users actually use it, and, more importantly, how they want to use it. Feedback to the authors will encourage future improvements.

Chapter 10 introduces students to the Symbolic Toolbox. Although this interface to the symbolic computing power of Maple can perform extraordinary tasks, we concentrate solely on its performance in the solution of ODEs and systems of ODEs. This material is elementary, and can be studied well before its appearance in the manual would indicate.

The linear algebra required to master the use of MATLAB to solve linear systems of equations is discussed in Chapter 11. Many ODE books assume that students know this important topic, an assumption which is often unwarranted. For that reason the discussion is fairly complete.

The material learned in Chapter 11 is used in Chapter 12 for the solution of homogeneous, linear systems of ODEs with constant coefficients. The ease with which MATLAB does matrix algebra allows students to experience a wider variety of systems than they would without the use of MATLAB. We discuss the exponential matrix and its use in the solution of linear systems.

Finally, in Chapter 13 we return to `pplane6` and discuss its advanced uses in studying non-linear, planar, autonomous systems. We discuss nullclines, equilibrium points, the Jacobian matrix and linearization, the classification of equilibrium points, separatrices, and basins of attraction. The limit set of a solution is introduced and the Poincarè-Bendixson alternatives are described.

The order of the chapters is close to the order of the topics in many ODE books. However, the prerequisites for the chapters are frequently minimal. Here is a list of the major dependencies:

$$1 \to 2 \to 4 \to 5 \qquad 1 \to 2 \to 4 \to 8 \to 9$$
$$1 \to 2 \to 11 \to 12 \qquad 1 \to 3 \to 4 \to 6$$
$$1 \to 2 \to 10 \qquad 1 \to 3 \to 7 \to 11 \to 12 \to 13$$

The notation $1 \to 3 \to 7$ means that Chapter 7 depends on Chapter 3, which in turn depend on Chapter 1. It is clear that the material can be covered in many orders. This having been said, it must be added that toward the end of many of the chapters there is more advanced material, especially in the exercises in Chapters 8, 12, and 13.

This edition contains a further increase in the number of exercises offered at the end of each chapter. We provide more routine problems, as well as harder problems, and activities that could be assigned as computer projects. Not all of the questions are directly computer related. Many of them require the student to use the computer to make a conjecture, and then to verify their conjecture analytically. Where a problem requires the student to use aspects of MATLAB which are not explained in the text, these are explained as part of the problem.

Instructors of ODE courses are increasingly assigning projects to their students. The web page http://online.redwoods.edu/instruct/darnold/DEProj/Index.htm is a source of ODE projects that have been used at the College of the Redwoods.

The special MATLAB functions dfield6, pplane6, and odesolve described herein, as well as the solvers eul, rk2, and rk4 are the work of the senior author, and are not distributed with MATLAB. We give precise instructions for obtaining this software in the Appendix to Chapter 3. The best source for the software is the webpage http://math.rice.edu/~dfield. There you will find versions which are compatible with all versions of MATLAB going back to 3.5. This webpage is the location where the most up to date versions will be found in the future. There is also a link to the java based versions of dfield and pplane. For the best compatibility with this document, you should use the versions of dfield6, pplane6, and odesolve which are meant to be used with MATLAB version 6.5 and later.

Over the past ten years, we have had conversations with many mathematicians about this manual and the associated software. We would like to mention Henry Edwards, Herman Gollwitzer, Larry Shampine, Bob Devaney, Bob Williams, Al Taylor, Marty Golubitsky, Bill Beckner, Charles Biles, Beverly West, Tom Beale, and Ted Scheick. To those whom we did not mention, we offer our apologies as well as our gratitude.

The manuscript was prepared using Yand Y TEX, using a modified set of macros originally written by Jim Carlson. Michael Spivak was always ready with an answer to our many questions about TEX.

The students at Rice and College of the Redwoods deserve our special gratitude. They suffered through years of experimentation, complaining only when appropriate. They were very active participants in the preparation of the manual in this form.

David Arnold   John C. Polking
Eureka, California   Houston, Texas

July 2003

1. Introduction to MATLAB

MATLAB is an interactive, numerical computation program. It has powerful built-in routines that enable a very wide variety of computations. It also has easy to use graphics commands that make the visualization of results immediately available. In some installations MATLAB will also have a Symbolic Toolbox which allows MATLAB to perform symbolic calculations as well as numerical calculations. In this chapter we will describe how MATLAB handles simple numerical expressions and mathematical formulas.

MATLAB is available on almost every computer system. Its interface is similar regardless of the system being used. We will assume that you have sufficient understanding of your computer to start up MATLAB and that you are now faced with a window on your computer which contains the MATLAB prompt[1], >>, and a cursor waiting for you to do something. This is called the MATLAB Command Window, and it is time to begin.

Numerical Expressions

In its most elementary use, MATLAB is an extremely powerful calculator, with many built-in functions, and a very large and easily accessible memory. Let's start at the very beginning. Suppose you want to calculate a number such as $12.3(48.5 + \frac{342}{39})$. You can accomplish this using MATLAB by entering `12.3*(48.5+342/39)`. Try it. You should get the following:

```
>> 12.3*(48.5+342/39)
ans =
   704.4115
```

Notice that what you enter into MATLAB does not differ greatly from what you would write on a piece of paper. The only changes from the algebra that you use every day are the different symbols used for the algebraic operations. These are standard in the computer world, and are made necessary by the unavailability of the standard symbols on a keyboard. Here is a partial list of symbols used in MATLAB.

+	addition	−	subtraction
*	multiplication	^	exponentiation
/	right division	\	left division

While + and − have their standard meanings, * is used to indicate multiplication. You will notice that division can be indicated in two ways. The fraction $\frac{2}{3}$ can be indicated in MATLAB as either 2/3 or as 3\2. These are referred to as right division and left division, respectively.

```
>> 2/3
ans =
    0.6667
>> 3\2
ans =
    0.6667
```

---

[1] In the narrative that follows, readers are expected to enter the text that appears after the command prompt (>>). You must press the **Enter** or **Return** key to execute the command.

Exponentiation is quite different in MATLAB; it has to be, since MATLAB has no way of entering superscripts. Consequently, the power 4^3 must be entered as 4^3.

```
>> 4^3
ans =
      64
```

The order in which MATLAB performs arithmetic operations is exactly that taught in high school algebra courses. Exponentiations are done first, followed by multiplications and divisions, and finally by additions and subtractions. The standard order of precedence of arithmetic operations can be changed by inserting parentheses. For example, the result of 12.3*(48.5+342)/39 is quite different than the similar expression we computed earlier, as you will discover if you try it.

MATLAB allows the assignment of numerical values to variable names. For example, if you enter

```
>> x = 3
x =
      3
```

then MATLAB will remember that x stands for 3 in subsequent computations. Therefore, computing 2.5*x will result in

```
>> 2.5*x
ans =
      7.5000
```

You can also assign names to the results of computations. For example,

```
>> y = (x+2)^3
y =
    125
```

will result in y being given the value $(3 + 2)^3 = 125$.

You will have noticed that if you do not assign a name for a computation, MATLAB will assign the default name ans to the result. This name can always be used to refer to the results of the previous computation. For example:

```
>> 2+3
ans =
      5

>> ans/5
ans =
      1
```

MATLAB has a few preassigned variables or constants. The constant $\pi = 3.14159...$ is given the name pi.

```
>> pi
ans =
      3.1416
```

The square root of -1 is i.

```
>> sqrt(-1)
ans =
        0 + 1.0000i
```

Engineers and physicists frequently use i to represent current, so they prefer to use j for the square root of -1. MATLAB is well aware of this preference.

```
>> j
ans =
        0 + 1.0000i
```

There is no symbol for e, the base of the natural logarithms, but this can be easily computed as `exp(1)`.

```
>> exp(1)
ans =
      2.7183
```

Mathematical Functions

There is a long list of mathematical functions that are built into MATLAB. Included are all of the functions that are standard in calculus courses.

Elementary Functions

abs(x)	The absolute value of x, i.e. $|x|$.
sqrt(x)	The square root of x, i.e. $\sqrt{x}$.
sign(x)	The signum of x, i.e. 0 if $x = 0$, -1 if $x < 0$, and $+1$ if $x > 0$.

The Trigonometric Functions

sin(x)	The sine of x, i.e. $\sin(x)$.
cos(x)	The cosine of x, i.e. $\cos(x)$.
tan(x)	The tangent of x, i.e. $\tan(x)$.
cot(x)	The cotangent of x, i.e. $\cot(x)$.
sec(x)	The secant of x, i.e. $\sec(x)$.
csc(x)	The cosecant of x, i.e. $\csc(x)$.

The Inverse Trigonometric Functions

asin(x)	The inverse sine of x, i.e. $\arcsin(x)$ or $\sin^{-1}(x)$.
acos(x)	The inverse cosine of x, i.e. $\arccos(x)$ or $\cos^{-1}(x)$.
atan(x)	The inverse tangent of x, i.e. $\arctan(x)$ or $\tan^{-1}(x)$.
acot(x)	The inverse cotangent of x, i.e. $\text{arccot}(x)$ or $\cot^{-1}(x)$.
asec(x)	The inverse secant of x, i.e. $\text{arcsec}(x)$ or $\sec^{-1}(x)$.
acsc(x)	The inverse cosecant of x, i.e. $\text{arccsc}(x)$ or $\csc^{-1}(x)$.

3

The Exponential and Logarithm Functions

`exp(x)`	The exponential of x, i.e. e^x.
`log(x)`	The natural logarithm of x, i.e. $\ln(x)$
`log10(x)`	The logarithm of x to base 10, i.e. $\log_{10}(x)$.

The Hyperbolic Functions

`sinh(x)`	The hyperbolic sine of x, i.e. $\sinh(x)$.
`cosh(x)`	The hyperbolic cosine of x, i.e. $\cosh(x)$.
`tanh(x)`	The hyperbolic tangent of x, i.e. $\tanh(x)$.
`coth(x)`	The hyperbolic cotangent of x, i.e. $\coth(x)$.
`sech(x)`	The hyperbolic secant of x, i.e. $\text{sech}(x)$.
`csch(x)`	The hyperbolic cosecant of x, i.e. $\text{csch}(x)$.

The Inverse Hyperbolic Functions

`asinh(x)`	The inverse hyperbolic sine of x, i.e. $\sinh^{-1}(x)$.
`acosh(x)`	The inverse hyperbolic cosine of x, i.e. $\cosh^{-1}(x)$.
`atanh(x)`	The inverse hyperbolic tangent of x, i.e. $\tanh^{-1}(x)$.
`acoth(x)`	The inverse hyperbolic cotangent of x, i.e. $\coth^{-1}(x)$.
`asech(x)`	The inverse hyperbolic secant of x, i.e. $\text{sech}^{-1}(x)$.
`acsch(x)`	The inverse hyperbolic cosecant of x, i.e. $\text{csch}^{-1}(x)$.

For a more extensive list of the functions available, see the MATLAB *User's Guide,* or MATLAB's online documentation.[2] All of these functions can be entered at the MATLAB prompt either alone or in combination. For example, to calculate $\sin(x) - \ln(\cos(x))$, where $x = 6$, we simply enter

```
>> x = 6
x =
     6

>> sin(x)-log(cos(x))
ans =
   -0.2388
```

Take special notice that $\ln(\cos(x))$ is entered as `log(cos(x))`. The function `log` is MATLAB's representation of the natural logarithm function.

Output Format

Up to now we have let MATLAB repeat everything that we enter at the prompt. Sometimes this is not useful, particularly when the output is pages in length. To prevent MATLAB from echoing what we

---

[2] MATLAB comes with extensive online help. Typing `helpdesk` at the MATLAB prompt should open MATLAB's helpdesk in a separate browser. You can also access MATLAB's standard help files by typing `help` at the MATLAB prompt. For a list of MATLAB's elementary functions, type `help elfun` at the MATLAB prompt.

tell it, simply enter a semicolon at the end of a command. For example, enter

```
>> q=7;
```

and then ask MATLAB what it thinks q is by entering

```
>> q
q =
     7
```

If you use MATLAB to compute $\cos(\pi)$, you get

```
>> cos(pi)
ans =
    -1
```

In this case MATLAB is smart enough to realize that the answer is an integer and it displays the answer in that form. However, cos(3) is not an integer, and MATLAB gives us -0.9900 as its value. Thus, if MATLAB is not sure that a number is an integer, it displays five significant figures in its answer. As another example, 1.57 is very close to $\pi/2$, and $\cos(\pi/2) = 0$. MATLAB gives us

```
>> cos(1.57)
ans =
    7.9633e-004
```

This is an example of MATLAB's exponential, or scientific notation. It stands for 7.9633×10^{-4}, or 0.00079633. In this case MATLAB again displays five significant digits in its answer. All of these illustrate the default format, which is called the short format. It is important to realize that although MATLAB only displays five significant digits in the default format, it is computing the answer to an accuracy of sixteen significant figures.

There are several other formats. We will discuss two of them. If it is necessary or desirable to have more significant digits displayed, enter format long at the MATLAB prompt. MATLAB will then display about sixteen significant digits. For example,

```
>> format long
>> cos(1.57)
ans =
    7.963267107332634e-004
```

There is another output format which we will find useful. If you enter format rat, then all numbers will be shown as rational numbers. This is called the rational format. If the numbers are actually irrational, MATLAB will find a very close rational approximation to the number.

```
>> cos(1.57)
ans =
    47/59021
```

The rational format is most useful when you are working with numbers you know to be rational. After using a different format, you can return to the standard, short format by entering format short.

Complex Arithmetic

One of the nicest features of MATLAB is that it works as easily with complex numbers as it does with real numbers. The complex number $z = 2 - 3i$ is entered exactly as it is written.

```
>> z = 2-3i
z =
    2.0000 - 3.0000i
```

Then if we enter $w = 3 + 5i$, we can calculate sums, products, and quotients of these numbers in exactly the same way we do for real numbers. For example,

```
>> w = 3+5i;
>> z*w
ans =
    21.0000 + 1.0000i
```

and

```
>> z/w
ans =
    -0.2647 - 0.5588i
```

Any of the arithmetic functions listed earlier can be applied to complex numbers. For example,

```
>> y = sqrt(w)
y =
    2.1013 + 1.1897i
```

and

```
>> y*y
ans =
    3.0000 + 5.0000i
```

Since $y^2 = w$, it is a square root of the complex number w. The reader might try `cos(w)` and `exp(w)`. In particular, the reader might wish to verify Euler's formula

$$e^{i\theta} = \cos(\theta) + i \sin(\theta)$$

for several values of θ, including $\theta = 2\pi, \pi, \pi/2$.

```
>> theta = pi; exp(i*theta), cos(theta) + i*sin(theta)
ans =
    -1.0000 + 0.0000i
ans =
    -1.0000 + 0.0000i
```

6

Note that several MATLAB commands can be placed on a single line, separated by commas, or semicolons, should you desire to suppress the output.

The ease with which MATLAB handles complex numbers has one drawback. There is at least one case where the answer is not the one we expect. Use MATLAB to calculate $(-1)^{1/3}$. Most people would expect the answer -1, but MATLAB gives us

```
>> (-1)^(1/3)
ans =
   0.5000 + 0.8660i
```

At first glance this may seem strange, but if you cube this complex number you do get -1. Consequently MATLAB is finding a *complex* cube root of -1, while we would expect a real root. The situation is even worse, since in most of the cases where this will arise in this manual, it is not the complex cube root we want. We will want the cube root of -1 to be -1.

However, this is a price we have to pay for other benefits. For MATLAB to be so flexible that it can calculate roots of arbitrary order of arbitrary complex numbers, it is necessary that it should give what seems like a strange answer for the cube root of negative numbers. In fact the same applies to any odd root of negative numbers. What we need is a way to work around the problem.

Notice that if $x < 0$, then $x = -1 \times |x|$, and we can find a negative cube root as $-1 \times |x|^{1/3}$. Here we are taking the real cube root of the positive number $|x|$, and MATLAB does that the way we want it done. But suppose the situation arises where we do not know beforehand whether x is positive or negative. What we want is

$$x^{1/3} = \begin{cases} |x|^{1/3}, & \text{if } x > 0; \\ 0, & \text{if } x = 0; \\ -1 \times |x|^{1/3}, & \text{if } x < 0. \end{cases}$$

To write this more succinctly we use the *signum* function $\text{sgn}(x)$ (in MATLAB it is denoted by $\text{sign}(x)$). This function is defined to be

$$\text{sgn}(x) = \begin{cases} 1, & \text{if } x > 0; \\ 0, & \text{if } x = 0; \\ -1, & \text{if } x < 0. \end{cases}$$

Thus, in all cases we have $x = \text{sgn}(x)\,|x|$, and the real cube root is

$$x^{1/3} = \text{sgn}(x)\,|x|^{1/3}.$$

In MATLAB, we would enter $\text{sign}(x)*\text{abs}(x)^{\wedge}(1/3)$.

Recording Your Work

It is frequently useful to be able to record what happens in a MATLAB session. For example, in the process of preparing a homework submission, it should not be necessary to copy all of the output from the computer screen. You ought to be able to do this automatically. The MATLAB `diary` command makes this possible.

For example, suppose you are doing your first homework assignment and you want to record what you are doing in MATLAB. To do this, choose a name, perhaps `hw1`, for the file in which you wish to record the output. Then enter `diary hw1` at the MATLAB prompt. From this point on, everything that appears in the Command Window will also be recorded in the file `hw1`. When you want to stop recording enter `diary off`. If you want to start recording again, enter `diary on`.

The file that is created is a simple text file. It can be opened by an editor or a word processing program and edited to remove extraneous material, or to add your comments. You can use the MATLAB editor for this process. To access it, first be sure you have stopped the editing process by executing `diary off`. Next, either click on the open file in the Toolbar, or select **Open file** from the **Edit** menu. The file selection window that opens is set to display only MATLAB files, so the text diary file will not be visible. Click on the Files of type: popup menu and select All Files. Then select the name of your diary file. You will now be able to make any changes you want in the file. You will also be able to print the file to get a hard copy.

To open your diary file, or any other file you create with MATLAB, you will have to know where it is stored on your computer. This means that you will have to understand the directory structure on your computer. (We are using directory as a synonym of file folder.) Your diary file is saved in the current directory. In MATLAB 6, the current directory is displayed in a small box at the top of the command window. Next to it there is a button with three dots (an ellipsis) on it. Clicking this button will open a new window containing the directory tree of your computer. You can make any directory the current directory by selecting it from the directory tree.

Different operating systems provide different ways of handling files and directories. However, it is possible to do a lot of file handling entirely within MATLAB. For this purpose MATLAB uses a combination of UNIX and DOS commands. You can find the name of the current directory by using the command `pwd`. The response will be the full address of the current directory in the language of your operating system. You can obtain a list of the files in the current directory with the commands `ls` or `dir`. You can change directories with the commands `cd` or `chdir`. You can make a new directory with the command `mkdir`. We suggest that you use the `help` command[3] to find out more, and experiment with these commands to learn what they do in your system.

Exercises

1. Use the standard procedure on your computer system to create a folder named `mywork`. In MATLAB change to this folder either by using the command `cd` at the MATLAB prompt or by clicking the ellipsis (...) button next to the Current Directory edit box on your MATLAB toolbar and browsing to the folder `mywork`. Look at the Current Directory edit box to be sure that `mywork` is the current directory. You can also use the command `pwd` at the MATLAB prompt. Clear the command window with the command `clc`, then clear your workspace of all variables with the command `clear`. Start a diary session with `diary hmwk1`. Read Chapter 1 (Introduction to Matlab) of this manual again, but this time enter each of the commands in the narrative at the MATLAB prompt as you read. When you are finished, enter the command `diary off`. Open the file hmwk1 in your favorite editor or word processor (or open it in Matlab's editor by typing `edit hmwk1` at the MATLAB prompt). Edit and correct any mistakes that you made. Save and print the edited file and submit the result to your instructor.

---

[3] For example, type `help cd` to obtain help on using the command `cd`.

2.   Plotting in MATLAB

MATLAB provides several methods for plotting the graphs of functions and more general curves. The easiest to use is what we will call EZ plotting, since it uses the command `ezplot` and its variants. While it is easy to use it lacks flexibility. We will briefly explain EZ plotting in the first section. The second method we will describe uses the commands `plot` and `plot3`. It provides more flexibility at the cost of some ease of use. It is the principal method used in this manual, so we will explain it in some detail. At the end of this chapter we will introduce MATLAB's handle graphics. The use of handle graphics gives the user complete control over the graphic, but it is not so easily used.

EZ plotting

To plot the sine function in MATLAB, simply execute the command

```
>> ezplot('sin(x)')
```

The result is shown in Figure 2.1. You will notice that `ezplot` produces the plot of $\sin(x)$ over the interval $[-2\pi, 2\pi]$. Next execute

```
>> ezplot('x*exp(-x^2)')
```

This time the plot, shown in Figure 2.2, is over an interval slightly bigger than $[-2.5, 2.5]$.

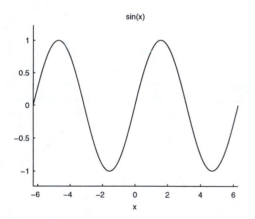

Figure 2.1. Plot of $\sin x$.

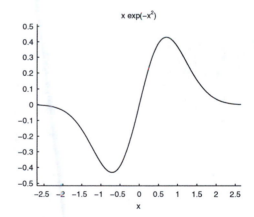

Figure 2.2. Plot of xe^{-x^2}.

Notice that you did not have to specify a plot interval. The command `ezplot` uses $[-2\pi, 2\pi]$ as the default interval over which to produce a plot. If the function is almost constant near the endpoints of the interval $[-2\pi, 2\pi]$, as is xe^{-x^2}, then ezplot chooses a smaller interval. If the default interval does not please you, you can choose your own. For example, execute

```
>> ezplot('sin(x)',[0,8*pi])
```

to plot the sine function over the interval $[0, 8\pi]$.

Example 1. *The initial value problem $y' = y + t$ with $y(0) = 0$ has the solution $y(t) = e^t - t - 1$. Plot this solution over the interval $0 \le t \le 2$.*

This can be done with the single command ezplot('exp(t)-t-1',[0,2]).

The command ezplot can also plot curves that are defined implicitly or parametrically. To learn about these features execute help ezplot. This command also illustrates the use of the command help. Whenever you run across a command you do not understand, use the help command. It will provide you with the information you need.

Example 2. *The solution to the initial value problem $y' = -2t/(3y^2 + 2y + 1)$ with $y(0) = 0$ satisfies the implicit relationship $y^3 + y^2 + y + t^2 = 0$. Plot the solution over the interval $-2 \le t \le 2$.*

If you execute the command ezplot('y^3+y^2+y+t^2 = 0'), you will get the plot of the solution shown in Figure 2.3. Notice that by default the solution is plotted over $-2\pi \le x \le 2\pi$. The important part of the solution would be better portrayed if the t-axis were reduced to $-2 \le t \le 2$ and the y-axis were reduced to $-2 \le y \le 1$. This can be done with the command ezplot('y^3+y^2+y+t^2',[-2, 2, -2, 1]). The result is shown in Figure 2.4. It is very typical that our first attempt is not completely satisfactory. Do not be afraid to redo a plot to make it look better.

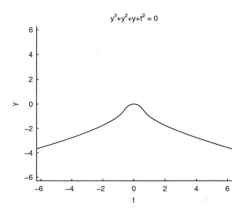

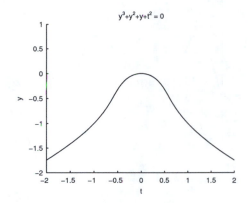

Figure 2.3. The solution in Example 2.

Figure 2.4. The solution in a smaller plot area.

The vector [-2, 2, -2, 1] that appears in the ezplot command means that the plot area will be limited by $-2 \le t \le 2$ and $-2 \le y \le 1$. This vector is referred to as the *axis*. You can discover the axis of the current plot by executing `axis` on the command line. Execute `help axis` to learn more about this versatile command. For example, the same reduction of plot area can be accomplished with the command `axis([-2, 2, -2, 1])` after the plot in Figure 2.3 was done.

If you own the Symbolic Toolbox, an add-on product to MATLAB, the command `ezplot` provides an excellent way to get a quick plot of the solution of an initial value problem.

Example 3. *Graph the solution of* $y' - y = e^{-t}\cos 3t$, *with* $y(0) = 1$.

Enter the command

```
>> dsolve('Dy - y = exp(-t)*cos(3*t)','y(0) = 1')
ans =
-2/13*exp(-t)*cos(3*t)+3/13*sin(3*t)*exp(-t)+15/13*exp(t)
```

Then the command

```
>> ezplot(ans)
```

will provide a plot of the solution. We discuss the Symbolic Toolbox in Chapter 10.

Matrices and Vectors in MATLAB

To use the `plot` command effectively it is necessary to know a little about how MATLAB works. A powerful feature of MATLAB is that every numerical quantity is considered to be a complex matrix![1] For those of you who do not already know, a *matrix* is a rectangular array of numbers. For example,

$$A = \begin{bmatrix} 1 & \pi & \sqrt{-1} \\ \sqrt{2} & 4 & 0 \end{bmatrix}$$

is a matrix with 2 rows and 3 columns.

If you want to enter the matrix A into MATLAB proceed as follows:

```
>> A = [1,pi,sqrt(-1);sqrt(2),4,0]
A =
   1.0000              3.1416              0 + 1.0000i
   1.4142              4.0000              0
```

Note that commas are used to separate the individual elements in a row, and semicolons are used to separate the rows of the matrix. You can also use *spaces* to separate (delimit) the entries in a row. Thus, the command `A = [1 pi sqrt(-1);sqrt(2) 4 0]` can be used to enter A into MATLAB.

The *size* of a matrix is the number of rows and columns. For example,

```
>> size(A)
ans =
      2      3
```

---

[1] A matrix whose entries are complex numbers.

verifies that A has 2 rows and 3 columns. Two matrices are said to have the same size if they have the same number of rows and the same number of columns.

Even single numbers in MATLAB are matrices. For example,

```
>> a = 5;
>> size(a)
ans =
        1        1
```

shows that MATLAB thinks that 5, or any other complex number, is a matrix with one row and one column.

A *vector* is a list of numbers.[2] It can be a vertical list, in which case it is called a *column vector*, or it can be a horizontal list, in which case it is called a *row vector*. Vectors are special cases of matrices, so you can enter the row vector $v = [1, -5, \pi, \sqrt{-1}]$ into MATLAB using the command

```
v = [1,-5,pi,sqrt(-1)]
```

or v = [1 -5 pi sqrt(-1)]. On the other hand, we can define a column vector with the command u = [1;2;3;4]. It is important to remember that MATLAB distinquishes between row and column vectors.

In MATLAB, the *length* of a vector is the number of elements in the list. For example, the MATLAB length of each of the vectors u and v defined in the previous paragraph is 4. The MATLAB command length will disclose the length of any vector. Try length(u) and length(v). Notice that u and v have the same length, but not the same size, since u is a column vector and *v* is a row vector.

This notion of length is not to be confused with the *geometric length* of a vector, defined to be the square root of the sum of the squares of the absolute values of the entries. MATLAB's command for finding the geometric length of a vector is norm. For example,

```
>> norm(v)
ans =
        6.0720
```

Addition, Subtraction, and Multiplication by Scalars

If *A* and *B* are matrices of the same size, then they can be added together. For example,

---

[2] The word *vector* is one of the most over used terms in mathematics and its applications. To a physicist or a geometer, a vector is a directed line segment. To an algebraist or to many engineers, a vector is a list of numbers. To users of more advanced parts of linear algebra, a vector is an element of a vector space. In this latter, most general case, a vector could be any of the above examples, a polynomial, a more general function, or an example of, quite literally, any class of mathematical objects which can be added together and scaled by multiplication.

All too often the meaning in any particular situation is not explained. The result is very confusing to the student. When the word vector appears, a student should make a concerted effort to discover the meaning that is used in the current setting.

When using MATLAB, the situation is clear. A vector is a list of numbers, which may be complex.

```
>> A = [1 2;3 4], B = [5,6;7,8], C = A + B
A =
     1     2
     3     4
B =
     5     6
     7     8
C =
     6     8
    10    12
```

You will notice that each element of the matrix C is sum of the corresponding elements in the matrices A and B. The same is true for the difference of two matrices. Try C-A, and see what you get (you should get B).

You can multiply or divide any matrix by a scalar.

```
>> v = ones(1,5); w = 4*v, z = v/2
w =
     4     4     4     4     4
z =
    0.5000    0.5000    0.5000    0.5000    0.5000
```

While multiplication and division by scalars are standard parts of matrix algebra, the addition of a scalar to, or the subtraction of a scalar from a matrix are not standard algebra. However, they are allowed in MATLAB. Try

```
>> m = v - 3, a = v + 4
m =
    -2    -2    -2    -2    -2
a =
     5     5     5     5     5
```

MATLAB's *transpose* operator (a single apostrophe) changes a column vector into a row vector (and vice-versa).

```
>> u = [1;2], v = u'
u =
     1
     2
v =
     1     2
```

Actually, . ' is MATLAB's transpose operator and ' is MATLAB's *conjugate transpose* operator. Enter A = [1+i,-2;3i,2-i], then type A ' and view the results. Note that rows of A have become columns, but each entry has been replaced with its complex conjugate. Type A . ' to appreciate the difference. If each entry of a matrix is a real number, then it doesn't matter whether you use ' or . '.

Array Operations and Array Smart Functions

MATLAB has built-in, element-by-element, operations for other mathematical operations on matrices. For example, should you need to multiply two vectors on an element-by-element basis, use MATLAB's `.*` operator.

```
>> v = [1,2,3,4], w = [5,6,7,8], u = v.*w
v =
     1 2 3 4
w =
     5 6 7 8
u =
     5 12 21 32
```

If you look closely, you will see that u is a vector of the same size as v and w, and that each element in u is the product of the corresponding elements in v and w. This operation is called *array multiplication*, and MATLAB's symbol for it is `.*`, not `*`. MATLAB uses `*` for matrix multiplication, which is quite different from array multiplication. Try entering `v*w` and see what happens. We will explain matrix multiplication in Chapter 11.

There are other array operations. All of them act element-by-element. Try `v./w` and `w./v`. This is *array right division*. Then try `v.\w` and `w.\v`, and compare the results. This is called *array left division*. There is one other array operation — *array exponentiation*. This is also an element-by-element operation. The operation `A.^2` results in every element of the matrix A being raised to the second power. For example, if `A = [1,2;3,4]`, the command

```
>> A = [1,2;3,4], B = A.^2
A =
     1     2
     3     4
B =
     1     4
     9    16
```

raises each entry in A to the second power. The command `A^2` is equivalent to `A*A`, which gives an entirely different result. Try it and see. For all array operations it is required that the matrices have exactly the same size. You might try `[1,2;3,4].*[1;1]` to see what happens when they are not.

The built-in MATLAB functions, which we discussed briefly in Chapter 1, are all designed to be *array smart*. This means that if you apply them to a matrix, the result will be the matrix obtained by applying the function to each individual element. For example:

```
>> theta = [0,pi/2,pi,3*pi/2,2*pi], y = cos(theta)
theta =
          0    1.5708    3.1416    4.7124    6.2832
y =
     1.0000    0.0000   -1.0000    0.0000    1.0000
```

This is an extremely important feature of MATLAB, as you will discover in the next section.

Plotting in MATLAB Using the plot Command

None of these array operations would be important if it were not so easy to create and use vectors and matrices in MATLAB. Here is a typical situation. Suppose we want to define a vector that contains a large number of equally spaced points in an interval $[a, b]$. MATLAB's start:increment:finish construct allows you to generate equally spaced points with ease. For example, the command

```
>> t = 0:0.2:1
t =
          0    0.2000    0.4000    0.6000    0.8000    1.0000
```

generates numbers from 0 to 1 in increments of 0.2.[3] The command y = t.^3 will produce a vector y with 6 entries, each the cube of the corresponding entry in the vector t.

```
>> y = t.^3
y =
          0    0.0080    0.0640    0.2160    0.5120    1.0000
```

We can get a rudimentary plot of the function $y = t^3$ by plotting the entries of y versus the entries of t. MATLAB will do this for us. The command

```
>> plot(t,y)
```

will produce a plot of y versus t in the current figure window. If no figure window exists, then the command plot will create one for you. MATLAB plots the 6 ordered pairs (t, y) generated by the vectors t and y, connecting consecutive ordered pairs with line segments, to produce an plot similar to the that shown in Figure 2.5.

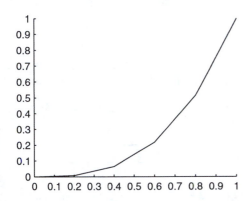

Figure 2.5. A simple plot

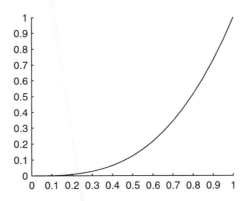

Figure 2.6. A simple plot with refined data.

---

[3] If you omit the increment, as in t = 0:10, MATLAB automatically increments by 1. For example, try q = 0:10.

Notice that the plot in Figure 2.5 is kinky. This is because we plotted too few points before MATLAB connected the dots with straight lines. If we use enough points we get a smooth looking curve. For example, the commands

```
>> t = 0:0.05:1; y = t.^3; plot(t,y), shg
```

produced the graph in Figure 2.6. This time there are 21 points, and the curve appears smooth. Notice that multiple commands can be entered on a single command line if they are separated with commas and/or semicolons. The command shg stands for "show the graph," and it brings the current figure window [4] to the front. It is a good idea to add shg to plot commands on the command window.

Example 4. *Use the* plot *command to graph* $f(x) = xe^{-x^2}$ *over the interval* $[-2, 2]$.

To accomplish this task, we need two MATLAB vectors. First, we need a vector x containing a large number of values between -2 and 2. We can do this with the command x = -2:0.1:2. Next we need a vector y containing the values of $f(x) = xe^{-x^2}$ at the points in x. This can be accomplished using the array operations. The operation .^ works element by element, so the vector x.^2 contains the squares of the values in x. Since MATLAB functions are array smart, exp(-x.^2) contains the values of e^{-x^2} for each of the entries of x. Finally, since .* is an array operation, x.*exp(-x.^2) contains the values of $f(x) = xe^{-x^2}$ for the entries in x. Thus the commands

```
>> x = -2:0.1:2;
>> y = x.*exp(-x.^2);
>> plot(x,y), shg
```

will produce the desired graph. Executing the command grid produces Figure 2.7.

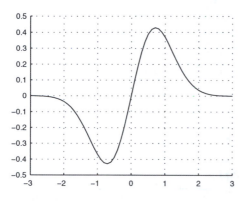

Figure 2.7. The graph for Example 4.

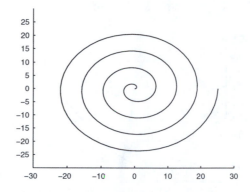

Figure 2.8. The parametric curve in Example 5.

---

[4] If several figure windows are open, the last one visited is the "current figure window."

Parametric plots. Notice that in Example 4 we used the command plot(x,y), where x and y were vectors of the same size. This command all by itself cares not where the two vectors came from. For any two vectors of the same size, the command will plot the (x, y) pairs and connect them with line segments. We can utilize this to produce parametric plots.

Example 5. *Plot the parametric curve defined by $t \rightarrow (t \cos t, t \sin t)$ for $0 \leq t \leq 8\pi$.*

We start with the command t = linspace(0,8*pi,200), which produces 200 equally spaced points[5] between 0 and 8π. Then x = t.*cos(t), and y = t.*sin(t) produce the corresponding values of the components of the desired curve. Finally, plot(x,y), shg produces the plot. To summarize, we use the commands

```
>> t = linspace(0,8*pi,200);
>> x = t.*cos(t);
>> y = t.*sin(t);
>> plot(x,y), shg
```

Curves in Three Dimensions. Three dimensional plots require the use of plot3 instead of plot, but otherwise the method is unchanged. The commands

```
>> t = linspace(0,20);
>> x = cos(t); y = sin(t); z = t;
>> plot3(x,y,z), shg
>> xlabel('x = cos t'); ylabel('y = sin t'); zlabel('t')
```

produce the helix in Figure 2.9. By default, the command linspace(a,b) produces a vector of 100 evenly spaced point between a and b.

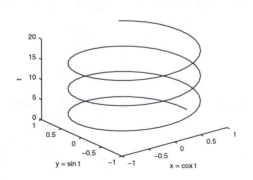

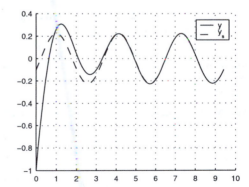

Figure 2.9. The spiral curve with $x = \cos t$, $y = \sin t$, and $z = t$.

Figure 2.10. Plots of y and y_S.

---

[5] Type help linspace for more information on using this command

Colors, markers, and line styles. Execute `help plot` at the MATLAB prompt and read the resulting help file. Pay particular attention to the various line styles, markers, and colors that can be used with MATLAB's `plot` command. You can produce strikingly different plots by varying the choice of these three attributes. Try the command `plot(t,y,'rx:')`, `shg`, and examine its affect on your plot. This plot command has the third argument `'rx:'`. The three symbols between the single quotes select, in order, a color, a marker, and a line style. Experiment with other combinations, such as `plot(t,y,'s')`, `plot(t,y,'md')`, and `plot(t,y,'k--')`. Use the `shg` command to view the results of each command.

Multiple graphs in a figure. There are several ways of doing this in MATLAB.

Example 6. *The initial value problem*

$$y'' + 2y' + 2y = \cos 2t, \quad \text{with} \quad y(0) = -1 \quad \text{and} \quad y'(0) = 2$$

has the solution $y(t) = y_t(t) + y_s(t)$, *where*

$$y_t(t) = e^{-t}\left(\frac{7}{10}\sin t - \frac{9}{10}\cos t\right), \quad \text{and} \quad y_s(t) = \frac{1}{5}\sin 2t - \frac{1}{10}\cos 2t$$

are the transient response and the steady-state solution, respectively. Plot the graphs of y and y_s *over the interval* $[0, 3\pi]$ *on the same figure. Use a solid line for y and a dashed line for* y_s. *Use a legend to label the graphs.*

We can use `t = linspace(0,3*pi)` to get 100 equally spaced t-values between 0 and 3π. The commands

```
>> y_t = exp(-t).*((7/10)*sin(t) -(9/10)*cos(t));
>> y_s = (1/5)*sin(2*t) - (1/10)*cos(2*t);
>> y = y_t + y_s;
```

compute the steady-state solution, the transient response, and the total response. A naive first try at plotting y and y_s together might be

```
>> plot(t,y)
>> plot(t,y_s,'--'), shg
```

However, MATLAB erases the first plot when the second plot command is executed. The commands

```
>> plot(t,y,t,y_s,'--')
>> grid on
>> legend('y','y_s')
```

will produce an image similar to that in Figure 2.10, with the graphs of both y and y_s on the same figure.

Notice that the parameters of the plot command come in groups. The first group consists of `t` and `y`, which is the data for the first curve. The second group has three entries, `t`, `y_s`, and `'--'`. This is the data for the the second curve plus a designation of a line style. The command `grid on` does just what it says — it adds a grid to the figure. Type `grid off` if you want to remove the grid. The `legend` command

18

produces the legend. It is only necessary to list the names in the order the graphs were produced. Notice that although entered as y_s, the label in the legend is subscripted.

A second solution to the problem of adding graphs to a figure involves the commands `hold on` and `hold off`. The command `hold on` tells MATLAB to add subsequent plots to the existing figure, without erasing what is already there. The command `hold off` tells MATLAB to return to the standard procedure of erasing everything before the next plot. This means that `hold on` is in effect until a `hold off` command is executed. Thus, we could have used the commands

```
>> plot(t,y)
>> hold on
>> plot(t,y_s,'--'), shg
>> hold off
```

to produce the plot in Figure 2.10.

A third way to plot two curves is to put the two sets of *y*-data into a matrix. The single command

```
>> plot(t,[y;y_s])
```

will cause the two curves to be plotted. The command `[y;y_s]` puts the two row vectors y and y_s into a matrix with two rows and as many columns as t. If A is a matrix with as many columns as t, then `plot(t,A)` will graph each row of A against t. A different color is automatically chosen for each curve.

Editing Graphics

MATLAB has extensive figure editing features. To use these make sure that the Figure Toolbar is visible by selecting **View**→**Figure Toolbar**.[6]

To change the appearance of a curve, click on the selection tool (the arrow pointing up and to the left in the Toolbar). Then click on the curve to select it. Now right-click[7] on the curve to bring up a context menu. There are several choices, but the ones we are most interested in allow us to change the line-width, line-style, and color of the curve. The use of these is amply clear once you try them once. If you select **Properties**, you are provided with a Property Editor window that gives complete control over the line properties, including adding markers to the data points and all of the available marker properties.

The Property Editor can also be accessed through the **Edit** menu. To edit the properties of a curve, first make that curve the *current object* by clicking on it. Then select **Edit**→**Current Object Properties ....** There are also Property Editors for the Figure (the area around the graph) and the Axes (the axis lines and labels, together with the background of the graph). In each case selection opens a new window which gives you complete control over the appearance of the object chosen.

For more information about editing your graphs, select **Help**→**Formatting Graphs** in any figure window.

---

[6] The notation **View**→**Figure Toolbar** means you should choose **Figure Toolbar** from the **View** menu located on the current figure window.

[7] Here we are assuming you have a mouse with two or more buttons. If you use a Macintosh and have only one button, use a control-click instead of a right-click.

Saving, Printing, and Exporting Your Plot

After learning how to plot graphs, you will want to print them. Simply use the print command in the File menu, or the print icon in the toolbar. **Edit→Print Setup** allows you to choose a printer and its properties and to choose between landscape or portrait output. **Edit→Page Setup** allows you to change many aspects of the printout.

To save a MATLAB figure to the clipboard, use the **Edit→Copy Figure** command. You will then be able to paste the figure into another document.

If you need to save a copy of a MATLAB figure to a graphics file, use the **File→Export** command. You will be given the opportunity to choose from a large variety of graphics formats. The choices made using **Edit→Page Setup** affect the exported file, so some flexibility is available.

It is possible to print what appears in the current figure window[8] to the default printer by simply entering `print` at the MATLAB prompt. In fact the `print` command provides many possibilities. Execute `help print` to explore them. For example, it is possible to export a figure to a graphics file using the `print` command at the command line. The command

```
>> print -deps junk.eps
```

will save the current figure as the encapsulated postscript file `junk.eps` in the current directory.

Script M-files

You will have noticed that producing a finished graphic in MATLAB requires the use of several commands entered at the command line. A finished graphic often requires several passes through these commands in order to get everything just right. Fortunately there is a way to get around the need to repeatedly enter a large list of commands as you improve on a graphic. It is possible to enter these commands into a text file, and execute all of them with one command.

Such files are called *M-files*. M-files come in two types, each with its own features. The type to use in building a complicated graphic is a *script M-file*, and we will describe them in this chapter. In addition there are *function M-files*, which can be used, for example, to extend MATLAB's library of functions. We will discuss function M-files in Chapter 4.

Let's start with a complicated graphing problem.

Example 7. *In one figure, plot the solutions to the differential equation $y' = y + t$ with initial conditions $y(0) = -2, -1, 0, 1, 2$ over the interval $0 \leq t \leq 2$.*

The general solution to the differential equation is $y(t) = Ce^t - t - 1$. The initial condition is $y(0) = C - 1$, so the constant satisfies $C = y(0) + 1$. The graphs can be drawn using the techniques we have already discussed, but it is easy to make mistakes in executing all of the commands needed. Instead we will create a script M-file. To see how this is done, choose the menu item **File–>New–>M-file**. The built-in MATLAB editor[9] will open at a blank page. You can also call up the editor by executing the

---

[8] MATLAB's `figure` command allows you to create multiple figure windows. If you have several figure windows open, click any figure window with your mouse to make it the current figure window.

[9] Starting with version 5.2, MATLAB has built-in editor on every platform. Of course, it is not

command `edit`. We will let MATLAB perform the simple arithmetic to compute the constant in addition to plotting the solutions. Enter the following list of commands into the blank editor page:

```
t = 0:0.05:2;
C = -2+1; plot(t,C*exp(t) - t - 1,'-')
hold on
C = -1+1; plot(t,C*exp(t) - t - 1,'-.')
C = 0+1; plot(t,C*exp(t) - t - 1,'--')
C = 1+1; plot(t,C*exp(t) - t - 1,'.')
C = 2+1; plot(t,C*exp(t) - t - 1,':')
grid on
xlabel('t')
ylabel('y')
title('Solutions to y'' = y + t.')
legend('y(0) = -2','y(0) = -1','y(0) = 0','y(0) = 1','y(0) = 2')
shg, hold off
```

Finally, save the file with a meaningful name such as `ch2examp7.m`. Now whenever you execute the command `ch2examp7` at the command line, all of these commands are executed and the figure appears. The figure is shown in Figure 2.11.

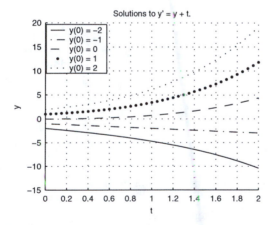

Figure 2.11. The family of solutions to a differential equation.

The commands `xlabel`, `ylabel`, and `title` at the end of the file `ch2examp7.m` are easily understood. You can get more information about them using the `help` command.

---

necessary to use the built-in editor. Any text editor will do. It is even possible to use a word processor, but if you do it is absolutely essential that you save the file as a text file.

Text strings in MATLAB

Notice that the labeling commands in `ch2examp7.m` accept strings of text as inputs. In MATLAB a string is any list of symbols entered between single quotes. This raises a question. How do we enter a string that contains a single quote, such as the title of the graph produced by `ch2examp7.m`? The answer is that we simply replace the single quote with a double quote, as in `'Solutions to y'' = y + t.'`.

We have now seen two different classes of data used by MATLAB. The first consists of numeric quantities like matrices and numbers. Since these are computed and stored to double the standard precision used in computers, the class is called *double*. The second class consists of strings of characters, and is called *char*. We will need to know how to change a number into a string. This is done with the command `num2str`. The use of this mnemonic command is illustrated by

```
>> x = 5.324, xstring = num2str(x)
x =
    5.3240
xstring =
5.324
```

So `x` is a number, and `xstring` is the same number transformed into a string. Notice that the only way to differentiate them on the command window is by the different indentation that MATLAB gives them. To get more information we can use the command

```
>> whos
  Name          Size                    Bytes  Class

  x             1x1                         8  double array
  xstring       1x5                        10  char array

Grand total is 6 elements using 18 bytes
```

This clearly illustrates the different classes of data. It is also possible to change a string into a number using `str2num`, but we will have less use for that.

Finally, we need to know how to concatenate strings to make bigger strings. An example is

```
>> string1 = 'Hello'; string2 = 'there';
>> string3 = [string1, ' ', string2,'.']
string3 =
Hello there.
```

The concatenated string, `string3`, is formed by formed by placing the four strings `string1`, `string2`, `' '`, and `'.'` between square brackets, separated by commas. The string `' '` provides the space between `string1` and `string2`, while `'.'` provides the period at the end.

A Little Programming

In the script M-file `ch2examp7.m` that we created for Example 7 we had to insert a line for each curve. We can replace that file with this one, which we will call `ch2examp7_2.m`.

```
t = 0:0.05:2;
Y = [];
for k = -2:2
  C = k+1; Y = [Y;C*exp(t)-t-1];
end
plot(t,Y)
grid on
xlabel('t'); ylabel('y')
title('Solutions to y'' = y + t.')
shg
```

The command `Y = []` introduces an empty matrix. Then at each of the five steps in the `for` loop, corresponding to $k = -2, -1, 0, 1, 2$, we first compute the constant `C = k+1`, and then the vector `C*exp(t)-t-1`, which is added as a new row to the matrix Y. At the end Y has a row for each solution and the `plot` command plots each of the five rows versus t, choosing a distinctive color for each. The use of such a simple `for` loop will simplify M-files whenever repetitive computations are required.

Other Issues with M-files

Organizing your files. Once you start using M-files, they tend to accumulate. A little organization can help keep some order in what could easily become chaos. If you haven't done so already (see Chapter 1, Exercise 1), use the standard procedure on your operating system to create a folder or directory called mywork (or invent a name of your own) in which to store your files. Make this directory the current directory by using the command cd[10] in the MATLAB command window or by clicking the ellipsis (. . .) button next to the Current Directory edit box on the toolbar of the command window and browsing to the new directory. You can check your success by looking at the contents of Current Directory edit box, or by executing pwd in the command window. It is important to understand that MATLAB first searches the current directory for script files. Hence, if you save your script file in a particular folder, make that folder the "current directory" before attempting to execute the file or MATLAB will be unable to find it. Alternatively, you can also put the folder mywork on MATLAB's path, but we will have more to say about this later.

Notice that we named the M-file used in Example 7 `ch2examp7.m`. This mnemonic enables us to identify this file in the future as the file used in Example 7 in Chapter 2. You will find it useful to invent a file naming system of your own.

The MATLAB editor. While any editor can be used to create M-files, there are definite advantages to using MATLAB's built-in editor. We will mention two.

- After creating your file, it can be saved and executed by selecting the menu item **Debug→Run** in the editor. You can use the accelerator key F5 to do the same thing after any changes you make.

---

[10] Type help cd to learn more about changing the "current directory."

- It is a rare thing to create an M-file that has no errors the first time through. The beauty of the M-file approach is that these errors can be eliminated one by one. Should your file contain errors, you should hear a "beep" or "click" when you press the F5 button in MATLAB's editor. An error message will be reported in the MATLAB command window, often as a link. Clicking this link will take you directly to the line in the MATLAB editor where the error occurs. This is an extremely useful tool for debugging your function M-files.

Handle graphics.

We have seen how the Property Editors enable great flexibility in editing figures. If we want the same flexibility when constructing figures from the command line or using M-files we need to use *handle graphics*.

Example 8. *The output of a forced undamped harmonic oscillator is given by*

$$x(t) = 2 \sin \frac{t}{2} \sin \frac{23t}{2}.$$

Plot the solution over the interval $[-2\pi, 2\pi]$. *In addition plot the envelope* $\pm 2 \sin(t/2)$ *with a line width of 2 pixels, and in a distinctive color.*

The only difficult part of this is increasing the line width. The normal line width is 0.4 pixels. Effecting the change using the Property Editor is easy, but we want to do this in an M-file, so that we can reproduce it easily. One way to do this is with these instructions.

```
t = linspace(-2*pi,2*pi,1000);
x = 2*sin(t/2).*sin(23*t/2);
env = 2*sin(t/2);
plot(t,x)
hold on
h = plot(t,[env;-env]);
set(h,'linewidth',2,'color','c');
hold off
axis([-2*pi,2*pi,-2.1,2.1])
xlabel('t')
ylabel('x(t)')
shg
```

We needed the very high resolution in `linspace` because the graph of *x* is highly oscillatory, and anything less was not sufficient. We inserted the `axis` command to better place the graphs in the figure.

However, the interesting commands are those that plot the envelope. Notice that the command h = plot(t,[env;-env]) outputs the quantity h. This is a vector of numerical handles to the two curves of the envelope. Execute h to see what the vector is. The next command, using `set`, changes the line width to 2 pixels and the color to cyan for both of the curves with handles in h. This is an example of the use of handle graphics. A black and white version of the result is shown in Figure 2.12.

Notice how the `set` command allows you to change individual properties of a curve. If you execute set(h(1)) you will see a full list of the properties, together with possible settings. The command

24

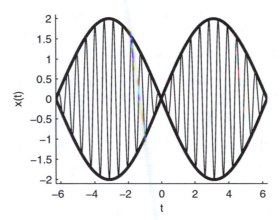

Figure 2.12. The phenomenon of beats.

`get(h(1))` will display a list of the settings for the curve with handle `h(1)`. Of course, you can replace `h(1)` with `h(2)` to see the results for the second curve. As is illustrated in Example 8, the `set` command can be used to change individual properties using `'property'`,`'property value'` pairs.

Curves are objects of the class *line*. There are a large number of object classes. An *axes* is the area on which a curve is plotted, including the axis lines and the associated markings. If you execute `gca`, you will see the handle of the current axes. Execute `set(gca)` to see the list of all of the properties of an axes. Any of these can be changed using the `set` command, just as we did for graphs. A *figure* is a figure window. You can find the handle of the current figure with the command `gcf`, and you can find a list of its properties using `set(gcf)`. If you want to find the handle of a graphics object, click the mouse on the object and execute `gco`. You can then find any or all of the current properties of that object with command `get(gco)`.

The easiest way to MATLAB's online documentation for a complete treatment of handle graphics is by selecting **Help→Graphics Help**. Find the link to the Handle Graphics Property Browser. This gives an overview of all of the classes of graphics objects and their properties. It is an invaluable aid if you use handle graphics.

Experiment further with the various properties of the objects in the figure you created in Example 8.

Exercises

In Exercises 1 – 6, find the solution to the indicated initial value problem, and use `ezplot` to plot it.

1. $y' = -ty$ with $y(0) = 1$ over $[0, 2]$.
2. $y' = t(y + 1)$ with $y(0) = 1$ over $[0, 2]$.
3. $y' = -y + \cos t$ with $y(0) = 2$ over $[0, 5]$.
4. $y' = -y^2 \cos t$ with $y(0) = 3$ over $[0, 3]$.
5. $x' = -3x + t + e^{-2t}$ with $x(0) = 0$ over $[0, 4]$.
6. $z' = 3z + t^2 e^{-t}$ with $z(0) = 1$ over $[0, 2]$.

25

In Exercises 7 – 10, the solutions are defined implicitly. Find the solution and plot it using `ezplot` in a region which displays the most important aspects of the solution. On the basis of your graph estimate the interval of existence.

7. $y' = (1 + 3t^2)/(3y^2 - 6)$ with $y(0) = 0$.   8. $(1 + 3y^2)y' = \cos(t)$ with $y(0) = 1$.

9. $y' = 3 \sin t/(3y^2 - 4)$ with $y(0) = 0$.   10. $y' = 3 \sin t/(3y^2 - 4)$ with $y(0) = -2$.

The `ezplot` command will also handle parametric equations. Try `ezplot('cos(t)','sin(t)',[0,2*pi])` to get a feel for how the command works. In Exercises 11 – 14, use the `ezplot` command to plot the parametric equations over the indicated time interval.

11. $x = \cos(2t) + 2\sin(2t)$, $y(t) = -\sin(2t)$, $[0, 2\pi]$

12. $x = \cos(t/2) + 2\sin(t/2)$, $y(t) = \sin(t/2)$, $[0, 4\pi]$

13. $x = e^{-t}(10\cos(5t) + 20\sin(5t))$, $y(t) = 10e^{-t}\sin(5t)$, $[0, 8\pi]$

14. $x = e^t(\cos(4t) - \sin(4t))$, $y(t) = 2e^t\sin(4t)$, $[0, 8\pi]$

If the Symbolic Toolbox is installed in your MATLAB system, use the `dsolve` command to find the solution of the first order initial value problems in Exercises 15 – 18. Use the `ezplot` command to plot the solution over the indicated interval.

15. $y' = -2ty$, $y(0) = 1$, $[-2, 2]$ 16. $y' + 2y = \cos(t)$, $y(0) = 1$, $[0, 20]$

17. $y' = 1 + y^2$, $y(0) = 1$, $[-\pi, \pi]$ 18. $y' + y/t = ty^2$, $y(2) = 3$, $[-4, 6]$

In Exercises 19 – 24, find the solution to the given initial value problem. Write a script M-file to plot the solution over the indicated interval properly annotated with labels and a title.

19. $y' + ty = y$ with $y(1) = 3$ over $[-2, 4]$

20. $ty' = 2y + t^3 \cos(t)$ with $y(\pi) = 0$ over $[-2\pi, 2\pi]$

21. $y' = y \sin(t)$ with $y(0) = 1$ over $[-2\pi, 2\pi]$

22. $y' = ty^3$ with $y(0) = -1$ over $[-3/4, 3/4]$

23. $y' + y \cos(t) = \cos(t)$ with $y(\pi) = 0$ over $[0, 4\pi]$

24. $y' = y \cos(t)$ with $y(0) = -1$ over $[0, 6\pi]$

25. On the same figure plot $y = \cos(x)$ and $z = \sin(x)$ over the interval $[0, 4\pi]$. Use different line styles or colors for each curve, and label the figure appropriately.

26. On the same figure plot the three curves $y = \sin(x)$, $y = x - x^3/6$, and $y = x - x^3/6 + x^5/120$ over the interval $[-3, 3]$. Use different line styles or colors for each curve, and label the figure appropriately. Do you recognize the relationship between these three functions?

27. On the same figure plot the graphs of the function $y = e^x$ and its Taylor approximations of order 1, 2, and 3 over the interval $[-3, 3]$. Use different line styles or colors for each curve, and label the figure appropriately.

28. Consider the functions $y_1 = x$, $y_2 = x^2$, and $y_3 = x^4$ on the interval $[0.1, 10]$. Plot these three functions on the same figure using the command `plot`. Now do the same thing with the other plotting commands `semilogx`, `semilogy`, and `loglog`. Turn in only the one that you think is most revealing about the relationship between these functions. Use different line styles or colors for each curve, and label the figure appropriately. (Plotting more than one curve on a figure using any of these commands follows the same procedure used with `plot`.)

For each set of parametric equations in Exercises 29 – 32, use a script file to create two plots. First, draw a plot of both x and y versus t. Use handle graphics to apply different linestyles and color to the plots of x and y, then add a legend, axis labels, and a title. Open a second figure window by placing the `figure` command at this point in your script, then draw a plot of y versus x. Add axis labels and a title to your second plot.

29. $x = \cos(t) - 3\sin(t)$, $y = -2\sin(t) - \cos(t)$, $[0, 6\pi]$

30. $x = \cos(2t) - 8\sin(2t)$, $y = -5\sin(2t) - \cos(2t)$, $[0, 4\pi]$

31. $x = e^{-t}(\cos(2t) + 3\sin(2t))$, $y = e^{-t}(7\sin(2t) - \cos(2t))$, $[-2\pi, 2\pi]$

32. $x = e^t(\cos(3t) - 3\sin(3t))$, $y = e^t(-2\sin(3t) - \cos(3t))$, $[-2\pi, 2\pi]$

33. In three dimensions plot the curve defined by

$$x = t\cos(t),$$
$$y = t\sin(t),$$
$$z = t,$$

over the interval $t \in [0, 4\pi]$ with the `plot3` command. Label the figure appropriately.

In Exercises 34 – 41, find the general solution of the differential equation. Then plot the family of solutions with the indicated initial values over the specified interval. We will use MATLAB notation to indicate the range of initial values. You can use the method of Example 7, but think about using a `for` loop.

34. $y' + y = \sin t$ on the interval $[0, 4\pi]$ with initial conditions $y(0) = -10 : 2 : 10$.

35. $y' + y = 5$ on the interval $[0, 4]$ with initial conditions $y(0) = -10 : 2 : 10$.

36. $y' + \cos(x) \cdot y = \cos(x)$ on the interval $[-10, 10]$ with initial conditions $y(0) = -10 : 2 : 10$.

37. $y' = y - 3e^{-t}$ on the interval $[-2, 2]$ with initial conditions $y(0) = -5 : 1 : 5$.

38. $y' = y\cos t - 3y$ on the interval $[0, 3]$ with initial conditions conditions $y(0) = -0.4 : 0.1 : 0.4$.

39. $y' = (1 + y^2)\cos t$ on the interval $[0, 4\pi]$ with initial $y(0) = -0.4 : 0.1 : 0.4$.

40. $2yy' = \cos t$ on the interval $[0, \pi]$ with initial conditions $y(\pi/2) = -3, -2, -1, 1, 2, 3$.

41. $(2 + 2y)y' = \sin t$ on the interval $[0, 4\pi]$ with initial conditions $y(0) = -3, -2, 0, 1, 2, 3$.

42. The voltage across the capacitor in a driven RC-circuit is modeled by the initial value problem $V_c' + V_c = \cos(t)$, $V_c(0) = 0$. The solution of the problem can be written $V_c = V_t + V_s$, where

$$V_s = \frac{1}{2}\cos(t) + \frac{1}{2}\sin(t) \quad \text{and} \quad V_t = -\frac{1}{2}e^{-t}.$$

The solution V_s is called the *steady-state solution* and the solution V_t is called the *transient solution*. On one plot, sketch the solutions V_s, V_t, and V_c in blue, red, and black, respectively, over the time interval $[0, 6\pi]$. Add a legend to your plot.

43. Use the appropriate sum-to-product identity from trigonometry to show that

$$\sin(12t) - \sin(14t) = -2\sin(t)\cos(13t).$$

On one plot, plot $y = -2\sin(t)\cos(13t)$ and its envelopes $y = \pm 2\sin(t)$ over the time interval $[-2\pi, 2\pi]$. Use the selection tool on the figure toolbar to select each envelope, then right-click the selected envelope to change both its color and linewidth.

44. Use the appropriate sum-to-product identity from trigonometry to show that

$$\cos(18t) - \cos(20t) = 2\sin(t)\sin(19t).$$

On one plot, plot $y = 2\sin(t)\sin(19t)$ and its envelopes $y = \pm 2\sin(t)$ over the time interval $[-2\pi, 2\pi]$. Using handle graphics in a script file, change the color and linewidth of the envelopes.

3.   Introduction to DFIELD6

A first order ordinary differential equation has the form

$$x' = f(t, x).$$

To solve this equation we must find a function $x(t)$ such that

$$x'(t) = f(t, x(t)), \qquad \text{for all } t.$$

This means that at every point $(t, x(t))$ on the graph of x, the graph must have slope equal to $f(t, x(t))$.

We can turn this interpretation around to give a geometric view of what a differential equation is, and what it means to solve the equation. At each point (t, x), the number $f(t, x)$ represents the slope of a solution curve through this point. Imagine, if you can, a small line segment attached to each point (t, x) with slope $f(t, x)$. This collection of lines is called a *direction line field*, and it provides the geometric interpretation of a differential equation. To find a solution we must find a curve in the plane which is tangent at each point to the direction line at that point.

Admittedly, it is difficult to visualize such a direction field. This is where the MATLAB routine dfield6 demonstrates its value.[1] Given a differential equation, it will plot the direction lines at a large number of points — enough so that the entire direction line field can be visualized by mentally interpolating between the field elements. This enables the user to get some geometric insight into the solutions of the equation.

Starting DFIELD6

To see dfield6 in action, enter dfield6 at the MATLAB prompt. After a short wait, a new window will appear with the label DFIELD6 Setup. Figure 3.1 shows how this window looks on a PC running Windows. The appearance will differ slightly depending on your computer, but the functionality will be the same on all machines.

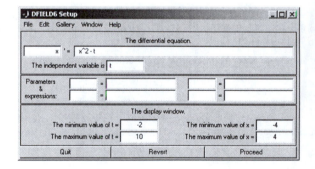

Figure 3.1. The setup window for dfield6.

---

[1] The MATLAB function dfield6 is not distributed with MATLAB. To discover if it is installed properly on your computer enter help dfield6 at the MATLAB prompt. If it is not installed, see the Appendix to this chapter for instructions on how to obtain it.

The DFIELD6 Setup window is an example of a MATLAB *figure window*. We have already seen figure windows in Chapter 2, but this one looks very different, so we see that a figure window can assume a variety of forms. In a MATLAB session there will always be one command window open on your screen and perhaps a number of figure windows as well.

The equation $x' = x^2 - t$ is entered in the edit box entitled "The differential equation" of the DFIELD6 Setup window. There is also an edit box for the independent variable and several edit boxes are available for parameters. The default values in "The display window" limit the independent variable t to $-2 \leq t \leq 10$ and the dependent variable x to $-4 \leq x \leq 4$. At the bottom of the DFIELD6 Setup window there are three buttons labelled **Quit**, **Revert**, and **Proceed**.

The Direction Field

We will describe the setup window in detail later, but for now click the button with the label **Proceed**. After a few seconds another window will appear, this one labeled DFIELD6 Display. An example of this window is shown in Figure 3.2.

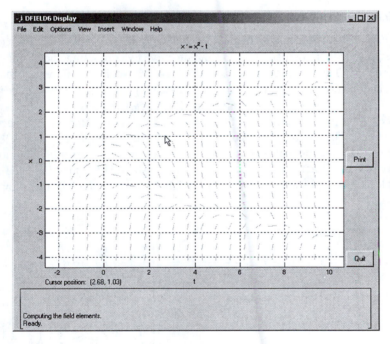

Figure 3.2. The display window for `dfield6`.

The most prominent feature of the DFIELD6 Display window is a rectangular grid labeled with the differential equation $x' = x^2 - t$ on the top, the independent variable t on the bottom, and the dependent variable x on the left. The dimensions of this rectangle are slightly larger than the rectangle specified in the DFIELD6 Setup window to accommodate the extra space needed by the direction field lines. Inside this rectangle is a grid of points, 20 in each direction, for a total of 400 points. At each such point with

29

coordinates (t, x) there is shown a small line segment centered at (t, x) with slope equal to $x^2 - t$. This collection of lines is a subset of the direction field.

There is a pair of buttons on the DFIELD6 Display window: **Quit** and **Print**. There are several menus: File, Edit, Options, Insert, and Help. Below the direction field there is a small window giving the coordinates of the cursor, and a larger message window through which `dfield6` will communicate with us. Note that the last line of this window contains the word "Ready," indicating that `dfield6` is ready to follow orders.

Initial Value Problems

The differential equation $x' = x^2 - t$ is in *normal form,* meaning that the derivative x' is expressed as a function of the independent variable t and the dependent variable x. You will notice from Figure 3.1 that `dfield6` requires the differential equation to be in normal form. Most differential equations have infinitely many solutions. In order to get a particular solution it is necessary to specify an initial condition. The differential equation with initial condition,

$$x' = f(t, x), \quad x(t_0) = x_0, \tag{3.1}$$

is called an *initial value problem.*

A *solution curve* of a differential equation $x' = f(t, x)$ is the graph of a function $x(t)$ which solves the differential equation. In particular we get a solution curve by computing and plotting the solution to an initial value problem. This is an easy process using `dfield6`. With the differential equation in normal form, we enter it and the other data in the setup window (see Figure 3.1). We then proceed to the display window (Figure 3.2). To solve with a given initial condition $x(t_0) = x_0$, we move the mouse to the point (t_0, x_0), using the cursor position display at the bottom of the figure to improve our accuracy, and then click the mouse button. The computer will compute and plot the solution through the selected point, first in the direction in which the independent variable is increasing (the "Forward" direction), and then in the opposite direction (the "Backward" direction). The result should be something like Figure 3.3. After computing and plotting several solutions, the display might look something like that shown in Figure 3.4.

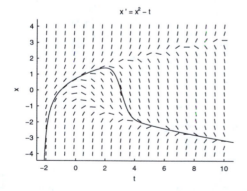

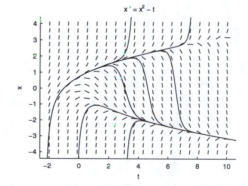

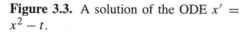

Figure 3.3. A solution of the ODE $x' = x^2 - t$.

Figure 3.4. Several solutions of the ODE $x' = x^2 - t$.

Finer Control of Data. The next example illustrates several features of `dfield6`, including how to be accurate with initial conditions.

Example 1. *The voltage y on the capacitor in a certain RC circuit is modeled by the differential equation* $y' + y = 3 + \cos x$, *where we are using the variable x to represent time. Use* `dfield6` *to plot the voltage over the interval* $0 \le x \le 20$, *assuming that* $y(0) = 1$.

You will notice that we are asked to solve the initial value problem

$$y' + y = 3 + \cos x, \quad y(0) = 1. \tag{3.2}$$

The dependent variable in this example is y and the independent variable is x. The differential equation $y' + y = 3 + \cos x$ is not in normal form, so we put it in normal form by solving the equation for y', getting $y' = -y + 3 + \cos x$.

Return to the DFIELD6 Setup window and select **Edit→Clear all**.[2] Notice that there are options on the Edit menu to clear particular regions of the DFIELD6 Setup window and each of these options possesses a keyboard accelerator. Enter the left and right sides of the differential equation $y' = -y + 3 + \cos x$, the independent variable (x in this case), and define the display window in the DFIELD6 Setup window as shown in Figure 3.5.[3]

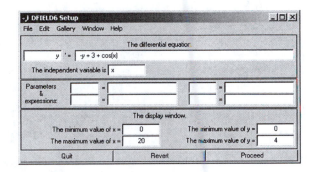

Figure 3.5. The setup window for $y' = -y + 3 + \cos x$.

Should your data entry become hopelessly mangled, click the **Revert** button to restore the original entries. The initial value problem in (3.2) contains no parameters, so leave the parameter fields in the DFIELD6 Setup window blank. Click the **Proceed** button to transfer the information in the DFIELD6 Setup window to the DFIELD6 Display window and start the computation of the direction field.

Choosing the initial point for the solution curve with the mouse is convenient, but it is often difficult to be accurate, even with the help of the cursor position display. Instead, in the DFIELD6 Display window, select **Options→Keyboard input**. Enter the initial condition, $y(0) = 1$, as shown in Figure 3.6. Click the **Compute** button in the DFIELD6 Keyboard input window to compute the trajectory shown in Figure 3.7.

---

[2] We continue to use the notation **Edit→Clear all** to signify that you should select "Clear all" from the Edit menu.

[3] MATLAB is case-sensitive. Thus, the variable Y is completely different from the variable y.

31

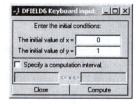

Figure 3.6. The initial condition $y(0) = 1$ starts the solution trajectory at $(0, 1)$.

Notice that it is not necessary to specify a computation interval. However, you can specify one if you wish by clicking the "Specify a computation interval" checkbox in the DFIELD6 Keyboard Input window (See Figure 3.6), and then filling in the starting and ending times of the desired solution interval. For example, start a solution trajectory with initial condition $y(0) = 0$, but set the computation interval so that $-\pi \le x \le \pi$. Try it!

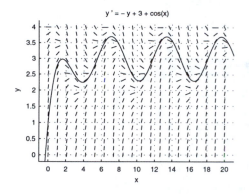

Figure 3.7. Solution of $y' + y = 3 + \cos x$, $y(0) = 1$.

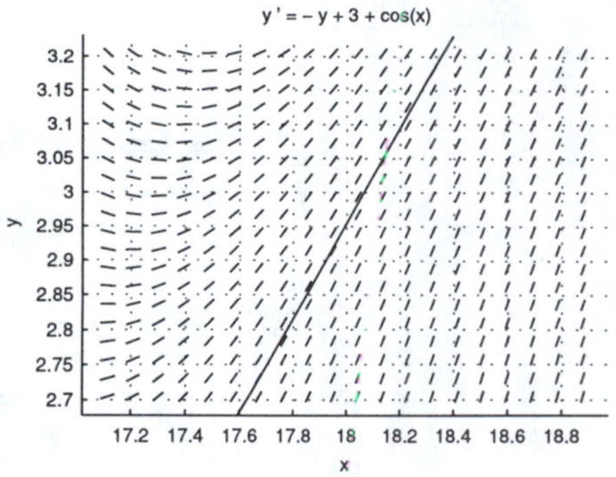

Figure 3.8. Zooming in to find $y(18)$.

Example 2. *For the voltage $y(x)$ computed in Example 1 find $y(18)$, accurate to 2 decimal places.*

From the graph of the solution in Figure 3.7 we can see that the voltage $y(18)$ is approximately 3. However, this is not accurate enough. To get more accuracy, we will increase the resolution using the zoom tools in `dfield6`. Select **Edit→Zoom in** in the DFIELD6 Display window, then single-click the (left) mouse button in the DFIELD6 Display window near the point $(18, 3)$. Additional "zooms" require that you revisit **Edit→Zoom in** before clicking the mouse button to zoom. There is a faster way to zoom in that is platform dependent. If you have a mouse with more than one button, click the right mouse button at the zoom point (or control-click the left mouse button at the zoom point). On a Macintosh, option-click the mouse button at the zoom point.[4] After performing a couple of zooms (results may vary

---

[4] Mouse actions are platform dependent in `dfield6`. See the front and back covers of this manual for a summary of mouse actions on various platforms.

on your machine), greater resolution is obtained. When you reach a point similar to Figure 3.8, you can use the cursor position display to see that $y(18) \approx 2.96$.

Existence and Uniqueness

It would be comforting to know in advance that a solution of an initial value problem exists, especially if you are about to invest a lot of time and energy in an attempt to find a solution. A second (but no less important) question is uniqueness: is there only one solution? Or does the initial value problem have more than one solution? Fortunately, existence and uniqueness of solutions have been thoroughly examined and there is a beautiful theorem that we can use to answer these questions.

Theorem 1. *Suppose that the function $f(t, x)$ is defined in the rectangle R defined by $a \le t \le b$ and $c \le x \le d$. Suppose also that f and $\partial f / \partial x$ are both continuous in R. Then, given any point $(t_0, x_0) \in R$, there is one and only one function $x(t)$ defined for t in an interval containing t_0 such that $x(t_0) = x_0$ and $x' = f(t, x)$. Furthermore, the function $x(t)$ is defined both for $t > t_0$ and for $t < t_0$, at least until the graph of x leaves the rectangle R through one of its four edges.*[5]

Example 3. *Use* `dfield6` *to sketch the solution of the initial value problem*

$$x' = x^2, \quad x(0) = 1.$$

Set the display window so that $-2 \le t \le 3$ and $-4 \le x \le 4$.

Enter the differential equation $x' = x^2$, the independent variable t, and the display window ranges $-2 \le t \le 3$ and $-4 \le x \le 4$ in the DFIELD6 Setup window. Click **Proceed** to compute the direction field. Select **Options→Keyboard input** in the DFIELD6 Display window and enter the initial condition $x(0) = 1$ in the DFIELD6 Keyboard input window. If all goes well, you should produce an image similar to that in Figure 3.9.

The differential equation $x' = x^2$ is in the form $x' = f(t, x)$, with $f(t, x) = x^2$. In addition, $f(t, x) = x^2$ and $\partial f / \partial x = 2x$ are continuous on the rectangle R defined by $-2 \le t \le 3$ and $-4 \le x \le 4$. Therefore, Theorem 1 states that the initial value problem has a solution as shown in Figure 3.9, and that this solution is unique. Use the mouse to experiment. You will see that any other solution is parallel to the first one and does not pass through the point $(0, 1)$.

Theorem 1 does not make a definitive statement about the domain of a solution. For example, does the solution in Figure 3.9 exist for all t or does it reach positive infinity in a finite amount of time? This question cannot be answered by `dfield6` alone, although it can provide a hint. Go back to the Setup window and change the display window to $0 \le t \le 1.5$ and $0 \le x \le 40$ in order to focus on the solution in Figure 3.9 near $t = 1$. When we proceed to the display window and recompute the solution, we get the result shown in Figure 3.10. This seems to indicate that the solution becomes infinite near $t = 1$. To check this out, we solve the differential equation. The general solution is $x(t) = 1/(C - t)$, where C is an arbitrary constant. Substituting the initial condition $x(0) = 1$ into the equation $x(t) = 1/(C - t)$,

---

[5] The notation $\partial f / \partial x$ represents the *partial derivative of f with respect to x*. Suppose, for example, that $f(t, x) = x^2 - t$. To find $\partial f / \partial x$, think of t as a constant and differentiate with respect to x to obtain $\partial f / \partial x = 2x$. Similarly, to find $\partial f / \partial t$, think of x as a constant and differentiate with respect to t to obtain $\partial f / \partial t = -1$.

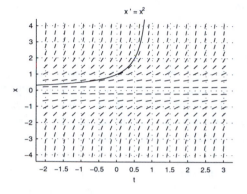

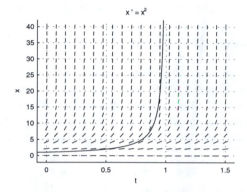

Figure 3.9. The solution of $dx/dt = x^2$, $x(0) = 1$ is unique.

Figure 3.10. The solution "blows up" at $t = 1$.

we find that $1 = 1/(C - 0)$, or $C = 1$, so the solution is $x(t) = 1/(1 - t)$. From this equation, we see that $\lim_{t \to 1^-} x(t) = +\infty$. Mathematicians like to say that the solution "blows up" at $t = 1$.[6] In this particular case, if the independent variable t represents time (in seconds), then the solution trajectory reaches positive infinity when one second of time elapses.

Example 4. *Consider the differential equation*

$$\frac{dx}{dt} = x^2 - t.$$

Sketch solutions with initial conditions $x(2) = 0$, $x(3) = 0$, and $x(4) = 0$. Determine whether or not these solution curves intersect in the display window defined by $-2 \le t \le 10$ and $-4 \le x \le 4$.

Go to the setup window and select **Gallery→default equation**. The correct data will be entered. Click **Proceed** to transfer this information and begin computation of the direction field in the DFIELD6 Display window. Select **Options→Keyboard input** in the DFIELD6 Display window and compute solutions for each of the initial conditions $x(2) = 0$, $x(3) = 0$, and $x(4) = 0$. If all goes well, you should produce an image similar to that in Figure 3.11.

The ODE $x' = x^2 - t$ is in normal form, $x' = f(t, x)$, with $f(t, x) = x^2 - t$. Both f and $\partial f/\partial x = 2x$ are continuous on the display window defined by $-2 \le t \le 10$ and $-4 \le x \le 4$. In Figure 3.11, it appears that the solution trajectories merge into one trajectory near the point $(6, -2.4)$ (or perhaps even sooner). However, Theorem 1 guarantees that solutions cannot cross or meet in the display window of Figure 3.11.

This situation can be analyzed by zooming in near the point $(6, -2.4)$. After performing numerous zooms, some separation in the trajectories begins to occur, as shown in Figure 3.12. Without Theorem 1, we might have mistakenly assumed that the trajectories merged into one trajectory.

It is also possible to zoom in by dragging a "zoom box". If you have a two button mouse, this can be done by depressing the right mouse button, then dragging the mouse. Once the zoom box is drawn

---

[6]   The graph of the solution reaches infinity (or negative infinity) in a finite time period.

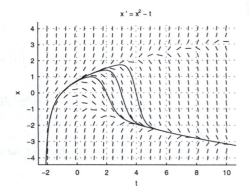

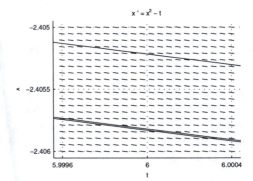

Figure 3.11. Do the trajectories intersect?

Figure 3.12. The trajectories don't merge or cross.

around the area of interest, release the mouse button and the contents of the zoom box will be magnified to the full size of the display window. The same effect can be achieved on a Macintosh with a one button mouse by depressing the option key while clicking and dragging.

Dfield6 allows you to "zoom back" to revisit any previously used window. Select **Edit→Zoom back** in the DFIELD6 Display window. This will open the DFIELD6 Zoom back dialog box pictured in Figure 3.13. Select the window you wish to revisit and click the **Zoom** button.

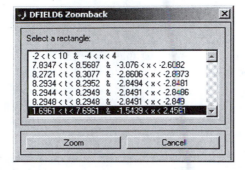

Figure 3.13. Select a zoom window and click the **Zoom** button.

Qualitative Analysis

Suppose that you model a population with a differential equation. If you want to use your model to predict the exact population in three years, then you will need to find an analytic or a numerical solution. However, if your only interest is what happens to the population after a long period of time, a qualitative approach might be easier and more appropriate.

Example 5. *Let $P(t)$ represent a population of bacteria at time t, measured in millions of bacteria.*

Suppose that P is governed by the logistic model

$$\frac{dP}{dt} = rP\left(1 - \frac{P}{K}\right). \tag{3.3}$$

Assume that $r = 0.75$ and $K = 10$ and suppose that the initial population at time $t = 0$ is $P(0) = 1$. What will happen to this population over a long period of time?

Let's first examine the model experimentally using dfield6. Instead of filling out the DFIELD6 Setup window by hand, we can use the gallery by choosing **Gallery→logistic equation**. Notice that the *parameters r* and *K* have been given the correct values. To provide more room below $P = 0$, set the minimum value of *P* to be −4, as shown in Figure 3.14. Click the **Proceed** button to start the computation of the direction field.

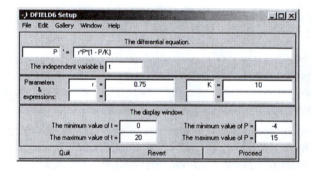

Figure 3.14. Setup window for $dP/dt = rP(1 - P/K)$.

Plot a few solutions by clicking the mouse at various points with $P > 0$. Notice that each of these solutions tends to 10 as t increases. Remember that $K = 10$. This is not a coincidence, and we will return to this point later. Some solution curves are shown in Figure 3.15.

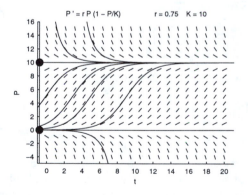

Figure 3.15. Solutions to $P' = rP(1 - P/K)$.

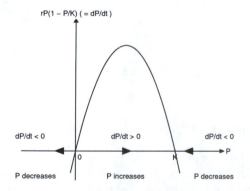

Figure 3.16. The plot of $rP(1 - P/K)$ versus P.

36

This behavior can be predicted quite easily using qualitative analysis. If you plot the right hand side of equation (3.3) versus P (i.e., plot $rP(1 - P/K)$ versus P), the result is the inverted parabola seen in Figure 3.16. Set $rP(1 - P/K)$ equal to zero to find that the graph crosses the P-axis in Figure 3.16 at $P = 0$ and $P = K$. These are called *equilibrium points*. It is easily verified that $P(t) = K$ is a solution of $dP/dt = rP(1 - P/K)$ by substituting $P(t) = K$ into each side of the differential equation and simplifying. Similarly, the solution $P(t) = 0$ is easily seen to satisfy the differential equation.

Although the solutions $P(t) = 0$ and $P(t) = K$ might be considered "trivial" since they are constant functions, they are by no means trivial in their importance. The solutions $P(t) = 0$ and $P(t) = K$ are called *equilibrium solutions*. For example, if $P(t) = K$, the growth rate dP/dt of the population is zero and the population remains at $P(t) = K$ forever. Similarly, if $P(t) = 0$, the growth rate dP/dt equals zero and the population remains at $P(t) = 0$ for all time.

When the graph of $rP(1 - P/K)$ (which is equal to dP/dt if P is a solution) falls below the P-axis in Figure 3.16, then $dP/dt < 0$ and the first derivative test implies that $P(t)$ is a decreasing function of t. On the other hand, when the graph of $rP(1 - P/K)$ rises above the P-axis in Figure 3.16, then $dP/dt > 0$ and $P(t)$ is an increasing function of t. These facts are summarized by the arrows on the P axis in Figure 3.16. This is an example of a *phase line*. The information on the phase line indicates that a population beginning between 0 and K million bacteria has to increase to the equilibrium value of K million bacteria. If the starting population is greater than K million then the population decreases to K million. For this reason the parameter K is called the *carrying capacity*.

Now let's go back to the DFIELD6 Display window. Select **Options→Keyboard input** and start solution trajectories with initial conditions $P(0) = 0$ and $P(0) = 10$. For the second trajectory you can enter $P = K$ instead of $P = 10$, since the use of parameters is allowed in the keyboard input window. In Figure 3.15, note that these equilibrium solutions are horizontal lines. Select **Options→Solution direction→Forward** and **Options→Show the phase line** in the DFIELD6 Display window. Dfield6 aligns the phase line from Figure 3.16 in a vertical direction at the left edge of the direction field in the DFIELD6 Display window. To see the motion along the phase line it is a good idea to slow the computations. Choose **Options→Solver settings** and move the slider to less than 10 solutions steps per second.

Next begin the solution with initial condition $P(0) = 1$ and note the action of the animated point on the phase line. As the solution trajectory in the direction field approaches the horizontal line $P = 10 = K$, the point on the phase line approaches equilibrium point $P = 10$ on the phase line, as shown in Figure 3.15. It would appear that a population with initial conditions and parameters described in the original problem statement will have to approach 10 million bacteria with the passage of time.

If you chose to slow the computation, perhaps you noticed something you had seen only fleetingly before. When a computation is started, a new button labelled **Stop** appears on the DFIELD6 Display window. If you click this button, the solution of the trajectory in the current direction is halted.

Experiment with some other initial conditions. Note that solutions beginning a little above or a little below the equilibrium solution $P = 10$ tend to move back toward this equilibrium solution with the passage of time. This is why the solution $P = 10$ is called an *assymptotically stable* equilibrium solution. However, solutions beginning a little above or a little below the equilibrium solution $P = 0$ tend to move away from this equilibrium solution with the passage of time. The solution $P = 0$ is called an *unstable* equilibrium solution. You can review the results of our experiments in Figure 3.15.

Using MATLAB While DFIELD6 is Open

All of the features of MATLAB are available while dfield6 is open. You can use MATLAB commands to plot to the DFIELD6 Display window, or you can open another figure window by typing figure at the MATLAB prompt. When more than one figure is open, it is important to remember that plotting commands are directed to the *current figure*. This is always the most recently visited window. You can make a particular figure active by clicking on it. If the DFIELD6 Setup window is the current figure, your plot command will be directed to it. It will be executed correctly, but it will not change the appearance of the window, so it will look as though nothing happened. This is an annoying outcome. When you have more than one figure window open, it is a good idea to click on the figure where you want a plot to be executed just before issuing the command.

Example 6. *The behavior of a population is modeled by the logistic equation*

$$P' = rP\left(1 - \frac{P}{K}\right),$$

with $r = 1$. However, in this case the carrying capacity is changing with time according to the equation $K = K(t) = 3 + t$. Use dfield6 *to plot several solutions. Plot the carrying capacity on the DFIELD6 Display window to facilitate comparison between the long-term behavior of the solutions and the carrying capacity.*

First we enter the equation into the DFIELD6 Setup window as in Figure 3.17. The only new feature here is that we entered the carrying capacity K = 3 + t as an expression. Any mathematical expression involving the dependent and independent variables can be entered.

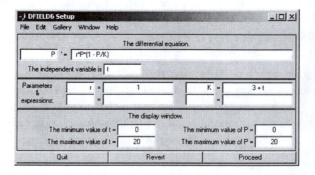

Figure 3.17. The setup window for the equation in Example 6.

Next we proceed to the DFIELD6 Display window and plot a few solutions (see Figure 3.18). Notice that ultimately all solutions seem to merge together and increase linearly, as does the carrying capacity. To see the relationship between the limiting behavior of the solutions and the carrying capacity more clearly, we use the commands

```
t = linspace(-2,22);
plot(t,3+t,'r')
```

38

to plot the carrying capacity in red. We clicked on the DFIELD6 Display window just before executing the `plot` command to make sure that it is the current figure. The result is shown in Figure 3.18, where we thickened the graph of the carrying capacity since we are not able to show a red curve in this manual. This can be done with the command `plot(t,3+t,'linewidth',2)`. You will notice that the limiting behavior of the solutions is linear growth, parallel to the graph of the carrying capacity.

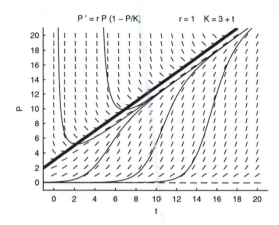

Figure 3.18. Plotting in the DFIELD6 Display window.

Subscripts and Greek Letters

The voltage V_c on the capacitor in an RC-circuit satisfies the differential equation

$$RCV_c' + V_c = A\cos\omega t, \tag{3.4}$$

where R is the resistance in ohms, C is the capacitance in farads, and $A\cos\omega t$ is a sinusoidal external voltage with amplitude A and frequency ω.

Example 7. *Use* `dfield6` *to study the response of an RC-circuit to external voltages of different frequencies. Use $R = 0.05\Omega$ and $C = 2F$. For $A = 5$ and $\omega = 1,\ 2,\ 5,\ 10,\ 20,\ 50,$ and 100 find the amplitude of the steady-state response with $V_c(0) = 0$. Why do you think an RC circuit is sometimes called a low-pass filter?*

Entering this equation into the DFIELD6 Setup window is easy since it is in the gallery. Choose **Gallery→RC circuit**, and make the needed change $R = 0.05$. The result is Figure 3.19. Now it is only necessary to change the input for the parameter ω and, when necessary, the maximum value of t to complete the exercise. In doing so you will notice that low frequency voltages pass through the RC circuit with their amplitudes practically unchanged, while high frequencies are attenuated. Hence the name *low-pass filter.* The results for two frequencies are shown in Figures 3.20 and 3.21.

Notice that the subscripted voltage V_c is entered as `V_c` into the DFIELD6 Setup window, and appears nicely subscripted in the DFIELD6 Display window. This is an example of TEX (or LATEX) notation. If you want a subscripted quantity to appear on a MATLAB figure window it is only necessary

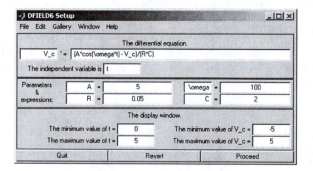

Figure 3.19. The setup window for Example 7.

to precede the subscript by an underscore. If the subscript contains more than one letter, put the entire subscript between curly brackets ({}). If you have a superscripted quantity, precede the superscript with a caret (^). Finally, notice that the frequency ω is entered in the setup window as \omega and appears in the display window in its proper Greek form. This, too, is TEX notation. Most Greek letters, including some upper case letters, can be treated this way. Simply spell out the name, preceded by a backslash. For example, you can use \alpha, \beta, \gamma, \theta, \phi, \Delta, \Omega, and \Theta.

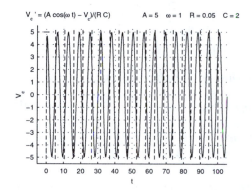

Figure 3.20. Response for $\omega = 1$.

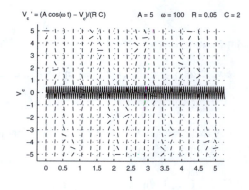

Figure 3.21. Response for $\omega = 100$.

Editing the Display Window

The appearance of the Display Window can be changed in a variety of ways.

Changing window settings. The menu item **Options→Windows settings** provides several ways to alter the appearance of the DFIELD6 Display window. Selecting this item will open the DFIELD6 Windows settings dialog box (see Figure 3.22).

The first option involves the three radio buttons, and allows you to choose between a line field, a vector field, or no field at all. Some people prefer to use a *vector field* to a direction line field. In a vector

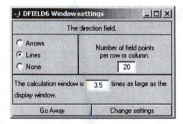

Figure 3.22. DFIELD6 Window settings.

field, a vector is attached to each point instead of the line segment used in a direction field. The vector has its base at the point in question, its direction is the slope, and the length of the vector reflects the magnitude of the derivative. Click the **Change settings** button to make any change you select.

There is an edit box in the DFIELD6 Window settings dialog that allows the user to choose the number of field points displayed. The default is 20 points in each row and in each column. Change this entry to 10, hit **Enter**, then click the **Change settings** button to note the affect on the direction field.

The design of `dfield6` includes the definition of two windows: the DFIELD6 Display window and the *calculation window*. When you start `dfield6`, the calculation window is 3.5 times as large as the display window in each dimension. The computation of a solution will stop only when the solution curve leaves the calculation window. This allows some room for zooming to a larger display window without having incomplete solution curves. It also allows for some reentrant solution curves — those which leave the display window and later return to it. The third item in the DFIELD6 Window settings dialog box controls the relative size of the calculation window. It can be given any value greater than or equal to 1. The smaller this number the more likely that reentrant solutions will be lost. The default value of 3.5 seems to meet most needs, but if you are losing too many reentrant solutions you can increase this parameter.

Marking initial points. It is possible to mark the points at which the computation of solutions is started. To do this, select **Options**→**Mark initial points**. To stop doing so, select the same option to uncheck it. Initial points that are already plotted can be erased with the command **Edit**→**Erase all marked initial points**.

Level curves. Sometimes it is useful to plot level sets of functions in the DFIELD6 Display window. In Example 6, instead of plotting the curve $P = K = 3 + t$ from the command line we could have plotted the level curve $P - K = 0$. This can be done using the command **Options**→**Plot level curves**. Complete the window as shown in Figure 3.23 and click Proceed. If you want to remove the level sets, select **Edit**→**Erase all level curves**.

Erasing objects. Sometimes when you are preparing a display window for printing, you plot a solution curve you wish were not there. In the **Edit** menu there are several commands which allow you to erase items in the DFIELD6 Display window. In addition to those we have already explained, there are **Edit**→ **Erase all solutions** and **Edit**→**Erase all graphics objects**, which are self explanatory. The last item, **Edit**→**Delete a graphics object**, is much more flexible. It will allow you to delete individual solution curves, as well as text items and graphs you have added to the window. Simply choose the option and select the object you wish to delete with the mouse.

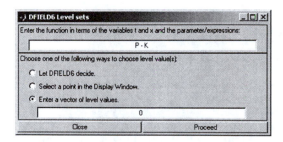

Figure 3.23. The DFIELD6 Level sets window.

Text objects in DFIELD6. The DFIELD6 Display window is a standard MATLAB figure window. Therefore all of the standard editing commands which we described in Chapter 2 are available. In particular the commands `xlabel`, `ylabel`, and `title` can be used to change these items. To add text at arbitrary points in the DFIELD6 Display window, select **Edit→Enter text on the Display Window**, enter the desired text in the Text entry dialog box, and then click the **OK** button. Use the mouse to click at the lower left point of the position in the figure window where you want the text to appear. It can easily happen that your placement of the text does not please you. If so, remove the text using **Edit→Delete a graphics object**, then try again.

Using the Property Editors with DFIELD6. We described the use of the Property Editors in Chapter 2. While these methods are very powerful and not difficult to use, they do not mix flawlessly with the interactive aspects of `dfield6`. You should be careful when you use them with the DFIELD6 Display window. It is a good idea to complete all of your `dfield6` work first, and only then begin to use the formatting commands. Do not mix them. It is not unusual that MATLAB freezes when the two are mixed. It is usually a good idea to select **Options→Make the display window inactive** before using the tools in the toolbar.

Other Features of DFIELD6

Printing, Saving, and Using Clipboards. You can print or export the DFIELD6 Display window in any of the ways described in Chapter 2. However, the easiest way to print the figure to the default printer is to click the **Print** button in the DFIELD6 Display window. The **Print** and **Quit** buttons and the message window will not be printed.

Saving and Loading DFIELD6 Equation and Gallery Files. Suppose that after entering all of the information into the DFIELD6 Setup window for Example 7, as it appears in figure 3.19, you decided to work on something else and come back to this example later. There are two ways to avoid the necessity of reentering the data. The first method is temporary. The menu option **Gallery→Add current equation to the gallery** will do just that, after prompting you for a name for the equation. When you are ready, you can choose this equation from the **Gallery** menu, and all of the data will be entered automatically.

However, if you have to quit `dfield6`, the new entry will no longer be there when you come back. For this situation you can use the command **File→Save the current equation ...**. This option allows you to record the information on the DFIELD6 Setup window in a file. Executing this option will bring up a standard file save menu, where you are given the option of saving the file with a filename and in a directory of your own choice. The file will be saved with the suffix `.dfs`. (It is not necessary to enter `.dfs`.) It can later be loaded back into `dfield6` using the command **File→Load an equation ...**.

It is also possible to save and load entire galleries using the appropriate commands on the **File** menu. Gallery files have the suffix `.dfg`. There is also a command that will delete the entire gallery, allowing you to start to build a gallery entirely your own, and another command that will reload the default gallery, if that is what you want.

Quitting DFIELD6. Always wait until the word "Ready" appears in the `dfield6` message window before you try to do anything else with `dfield6` or MATLAB. When you want to quit `dfield6`, the best way is to use the **Quit** buttons found on the DFIELD6 Setup or on the DFIELD6 Display windows. Either of these will close all of the `dfield6` windows in an orderly manner, and it will delete the temporary files that `dfield6` creates in order to do its business.

Plotting Several Solutions at Once. `Dfield6` allows you to plot several solutions at once. Select **Options→Plot several solutions** and note that the mouse cursor changes to "cross-hairs" when positioned over the direction field. Select several initial conditions for your solutions by clicking the mouse button at several different locations in the direction field. When you are finished selecting initial conditions, position the mouse cross-hairs over the direction field and press the **Enter** or **Return** key on your keyboard. Solution trajectories will emanate from the initial conditions you selected with the mouse.

Exercises

For the differential equations in Exercises 1–4, perform each of the following tasks.

a) Print out the direction field for the differential equation with the display window defined by $t \in [-5, 5]$ and $y \in [-5, 5]$. You might consider increasing the number of field points to 25 in the DFIELD6 Window settings dialog box. On this printout, sketch with a pencil as best you can the solution curves through the initial points $(t_0, y_0) = (0, 0)$, $(-2, 0)$, $(-3, 0)$, $(0, 1)$, and $(4, 0)$. Remember that the solution curves must be tangent to the direction lines at each point.

b) Use `dfield6` to plot the same solution curves to check your accuracy. Turn in both versions.

1. $y' = ty$.
2. $y' = y^2 - t^2$.
3. $y' = 2ty/(1 + y^2)$.
4. $y' = y(2 + y)(2 - y)$.
5. Use `dfield6` to plot a few solution curves to the equation $x' + x\sin(t) = \cos(t)$. Use the display window defined by $x \in (-10, 10)$ and $t \in (-10, 10)$.
6. Use `dfield6` to plot the solution curves for the equation $x' = 1 - t^2 + \sin(tx)$ with initial values $x = -3, -2, -1, 0, 1, 2, 3$ at $t = 0$. Find a display window which shows the most important features of the solutions by experimentation.

For the differential equations in Exercises 7–10 perform the following tasks.

a) Use `dfield6` to plot a few solutions with different initial points. Start with the display window bounded by $0 \le t \le 10$ and $-5 \le y \le 5$, and modify it to suit the problem. Print out the display window and turn it in as part of this assignment.

b) Make a conjecture about the limiting behavior of the solutions of as $t \to \infty$.

c) Find the general analytic solution to this equation.

d) Verify the conjecture you made in part b), or if you no longer believe it, make a new conjecture and verify that.

7. $y' + 4y = 8$.

8. $(1 + t^2)y' + 4ty = t$.

9. $ty' + ty = 2 - y$.

10. $(1 + t)y' = y(4 - y^2)$.

In Exercises 11–14 we will consider a certain lake which has a volume of $V = 100\,\text{km}^3$. It is fed by a river at a rate of $r_i\,\text{km}^3$/year, and there is another river which is fed by the lake at a rate which keeps the volume of the lake constant. In addition, there is a factory on the lake which introduces a pollutant into the lake at the rate of $p\,\text{km}^3$/year. This means that the rate of flow from the lake into the outlet river is $(p + r_i)\,\text{km}^3$/year. Let $x(t)$ denote the volume of the pollutant in the lake at time t, and let $c(t) = x(t)/V$ denote the concentration of the pollutant.

11. Show that, under the assumption of immediate and perfect mixing of the pollutant into the lake water, the concentration satisfies the differential equation $c' + ((p + r_i)/V)c = p/V$.

12. Suppose that $r_i = 50$, and $p = 2$.

    a) Assume that the factory starts operating at time $t = 0$, so that the initial concentration is 0. Use `dfield6` to plot the solution. Remember the definition of the concentration is x/V so you can be sure it is pretty small. Choose the dimensions of the display window carefully.

    b) It has been determined that a concentration of over 2% is hazardous for the fish in the lake. Approximately how long will it take until this concentration is reached? You can "zoom in" on the `dfield6` plot to enable a more accurate estimate.

    c) What is the limiting concentration? About how long does it take for the concentration to reach a concentration of 3.5%?

13. Suppose the factory stops operating at time $t = 0$, and that the concentration was 3.5% at that time. Approximately how long will it take before the concentration falls below 2%, and the lake is no longer hazardous for fish? Notice that $p = 0$ for this exercise.

14. Rivers do not flow at the same rate the year around. They tend to be full in the Spring when the snow melts, and to flow more slowly in the Fall. To take this into account, suppose the flow of our river is

$$r_i = 50 + 20\cos(2\pi(t - 1/3)).$$

Our river flows at its maximum rate one-third into the year, i.e., around the first of April, and at its minimum around the first of October.

    a) Setting $p = 2$, and using this flow rate, use `dfield6` to plot the concentration for several choices of initial concentration between 0% and 4%. (If your solution seems erratic, reduce the relative error tolerance using **Options**→**Solver settings**.) How would you describe the behavior of the concentration for large values of time?

    b) It might be expected that after settling into a steady state, the concentration would be greatest when the flow was smallest, around the first of October. At what time of the year does the highest concentration actually occur? Reduce the error tolerance until you get a solution curve smooth enough to make an estimate.

15. Use `dfield6` to plot several solutions to the equation $z' = (z - t)^{5/3}$. (**Hint:** Notice that when $z < t$, $z' < 0$, so the direction field should point down, and the solution curves should be decreasing. You might have difficulty getting the direction field and the solutions to look like that. If so read the section in Chapter 1 on complex arithmetic, especially the last couple of paragraphs.)

A differential equation of the form $dx/dt = f(x)$, whose right-hand side does not explicitly depend on the independent variable t, is called an *autonomous* differential equation. For example, the logistic model in Example 5 was autonomous. For the autonomous differential equations in Exercises 16 – 19, perform each of the following tasks. Note that the first three tasks are to be performed without the aid of technology.

    a) Set the right-hand side of the differential equation equal to zero and solve for the equilibrium points.

    b) Plot the graph of the right-hand side of each autonomous differential equation versus x, as in Figure 3.16. Draw the phase line below the graph and indicate where x is increasing or decreasing, as was done in Figure 3.16.

c) Use the information in parts a) and b) to draw sample solutions in the xt plane. Be sure to include the equilibrium solutions.

d) Check your results with dfield6. Again, be sure to include the equilibrium solutions.

e) If x_0 is an equilibrium point, i.e., if $f(x_0) = 0$, then $x(t) = x_0$ is an equilibrium solution. It can be shown that if $f'(x_0) < 0$, then every solution curve that has an initial value near x_0 converges to x_0 as $t \to \infty$. In this case x_0 is called a *stable* equilibrium point. If $f'(x_0) > 0$, then every solution curve that has an initial value near x_0 diverges away from x_0 as $t \to \infty$, and x_0 is called an *unstable* equilibrium point. If $f'(x_0) = 0$, no conclusion can be drawn about the behavior of solution curves. In this case the equilibrium point may fail to be either stable or unstable. Apply this test to each of the equilibrium points.

16. $x' = \cos(\pi x)$, $x \in [-3, 3]$.

17. $x' = x(x - 2)$, $-\infty < x < \infty$.

18. $x' = x(x - 2)^2$, $-\infty < x < \infty$.

19. $x' = x(x - 2)^3$, $-\infty < x < \infty$.

In Exercises 20 – 22 you will not be able to solve explicitly for all of the equilibrium points. Instead, turn the problem around. Use dfield6 to plot some solutions, and from that information calculate approximately where the equilibrium points are, and determine the type of each. In Exercise 20 you can check your estimate with the code:

```
f=inline('x*(1+exp(-x)-x^2)')
z=fzero(f,1)
f(z)
```

Similar methods will help for Exercises 21 and 22.

20. $x' = x(1 + e^{-x} - x^2)$, $-1 \le x \le 2$.

21. $x' = x^3 - 3x + 1$.

22. $x' = \cos x - 2x$.

The logistic equation $P' = r P(1 - P/K)$ is discussed in Examples 5 and 6. Usually the parameter's r and K are constants and in Example 5 we found that for any solution $P(t)$ which has a positive initial value we have $P(t) \to K$ as $t \to \infty$. For this reason K is called the *carrying capacity* of the system. However, in Example 6 we saw a case where the carrying capacity is not constant, yet we were able to show how the limiting behavior of the population related to the carrying capacity. In Exercises 23–26 you are to examine the long term behavior of solutions, especially in comparison to the carrying capacity. In particular:

a) Use dfield6 to plot several solutions to the equation. (It is up to you to find a display window that is appropriate to the problem at hand.)

b) Based on the plot done in part a), describe the long term behavior of the solutions to the equation. In particular, compare this long term behavior to that of K. It might be helpful to plot K on the display window as we did in Example 6. In the first case the solutions will all be asymptotic to a constant. In the other two the solutions will all have the same long term behavior. Describe that behavior in comparison to the graph of K. The results of Examples 5 and 6 should be helpful.

23. $K(t) = 1 - \frac{1}{2}e^{-t}$, $r = 1$. In this case $K(t)$ is monotone increasing, and $K(t)$ is asymptotic to 1. This might model a situation of a human population where, due to technological improvement, the availability of resources is increasing with time, although ultimately limited.

24. $K(t) = \sqrt{1 + t}$, $r = 1$. Again $K(t)$ is monotone increasing, but this time it is unbounded. This might model a situation of a human population where, due to technological improvement, the availability of resources is steadily increasing with time, and therefore the effects of competition are becoming less severe.

25. $K(t) = 1 - \frac{1}{2}\cos(2\pi t)$, $r = 1$. This is perhaps the most interesting case. Here the carrying capacity is periodic in time with period 1, which should be considered to be one year. This models a population of insects or small animals that are affected by the seasons. You will notice that the long term behavior as $t \to \infty$ reflects the behavior of K. The solution does not tend to a constant, but nevertheless all solutions have the same long term behavior for large values of t. In particular, you should take notice of the location of the

maxima and minima of K and of P and how they are related. You can use the "zoom in" option to get a better picture of this.

26. $K(t) = \sqrt{1+t} - \frac{1}{2}\cos(2\pi t)$, $r = 1$.

27. Despite the seeming generality of the uniqueness theorem, there are initial value problems which have more than one solution. Consider the differential equation $y' = \sqrt{|y|}$. Notice that $y(t) \equiv 0$ is a solution with the initial condition $y(0) = 0$. (Of course by $\sqrt{|y|}$ we mean the **nonnegative** square root.)

    a) This equation is separable. Use this to find a solution to the equation with the initial value $y(t_0) = 0$ assuming that $y \geq 0$. You should get the answer $y(t) = (t - t_0)^2/4$. Notice, however, that this is a solution only for $t \geq t_0$. Why?

    b) Show that the function

$$y(t) = \begin{cases} 0, & \text{if } t < t_0; \\ (t - t_0)^2/4, & \text{if } t \geq t_0; \end{cases}$$

    is continuous, has a continuous first derivative, and satisfies the differential equation $y' = \sqrt{|y|}$.

    c) For any $t_0 \geq 0$ the function defined in part b) satisfies the initial condition $y(0) = 0$. Why doesn't this violate the uniqueness part of the theorem?

    d) Find another solution to the initial value problem in a) by assuming that $y \leq 0$.

    e) You might be curious (as were the authors) about what `dfield6` will do with this equation. Find out. Use the rectangle defined by $-1 \leq t \leq 1$ and $-1 \leq y \leq 1$ and plot the solution of $y' = \sqrt{|y|}$ with initial value $y(0) = 0$. Also, plot the solution for $y(0) = 10^{-50}$ (the MATLAB notation for 10^{-50} is `1e-50`). Plot a few other solutions as well. Do you see evidence of the non-uniqueness observed in part c)?

An important aspect of differential equations is the dependence of solutions on initial conditions. There are two points to be made. First, we have a theorem which says that the solutions are continuous with respect to the initial conditions. More precisely,

Theorem. *Suppose that the function $f(t, x)$ is defined in the rectangle R defined by $a \leq t \leq b$ and $c \leq x \leq d$. Suppose also that f and $\partial f/\partial x$ are both continuous in R, and that*

$$\left|\frac{\partial f}{\partial x}\right| \leq L \quad \text{for all } (t, x) \in R.$$

If (t_0, x_0) and (t_0, y_0) are both in R, and if

$$\begin{aligned} x' &= f(t, x) \\ x(t_0) &= x_0 \end{aligned} \quad \text{and} \quad \begin{aligned} y' &= f(t, y) \\ y(t_0) &= y_0 \end{aligned}$$

then for $t > t_0$

$$|x(t) - y(t)| \leq e^{L(t-t_0)}|x_0 - y_0|$$

as long as both solution curves remain in R.

Roughly, the theorem says that if we have initial values that are sufficiently close to each other, the solutions will remain close, at least if we restrict our view to the rectangle R. Since it is easy to make measurement mistakes, and thereby get initial values off by a little, this is reassuring.

For the second point, we notice that although the dependence on the initial condition is continuous, the term $e^{L(t-t_0)}$ allows the solutions to get exponentially far apart as the interval between t and t_0 increases. That is, the solutions can still be extremely sensitive to the initial conditions, especially over long t intervals.

28. Consider the differential equation $x' = x(1 - x^2)$.

    a) Verify that $x(t) \equiv 0$ is the solution with initial value $x(0) = 0$.

b) Use `dfield6` to find approximately how close the initial value y_0 must be to 0 so that the solution $y(t)$ of our equation with that initial value satisfies $y(t) \leq 0.1$ for $0 \leq t \leq t_f$, with $t_f = 2$. You can use the display window $0 \leq t \leq 2$, and $0 \leq x \leq 0.1$, and experiment with initial values in the **Options→ Keyboard input** window, until you get close enough. Do not try to be too precise. Two significant figures is sufficient.

c) As the length of the t interval is increased, how close must y_0 be to 0 in order to insure the same accuracy? To find out, repeat part b) with $t_f = 4, 6, 8$, and 10.

The results of the last problem show that the solutions can be extremely sensitive to changes in the initial conditions. This sensitivity allows chaos to occur in deterministic systems, which is the subject of much current research.

One way to experience first hand the sensitivity to changes in the initial conditions is to try a little "target practice." For the ODEs in Exercises 29–33, use `dfield6` to find approximately the value of x_0 such that the solution $x(t)$ to the initial value problem with initial condition $x(0) = x_0$ satisfies $x(t_1) = x_1$. You should use the Keyboard input window to initiate the solution. Widen the window to allow a large number of digits in the edit window by clicking and dragging on the right edge. After an unsuccessful attempt try again with another initial condition. The Uniqueness Theorem should help you limit your choices. If you make sure that the Display Window is the current figure (by clicking on it), and execute `plot(t1,x1,'or')` at the command line, you will have a nice target to shoot at.

You will find that hitting the target gets more difficult in each of these problems. We allow you to "cheat" by starting a solution in the target, and finding the value at $t = 0$. However, be sure to try to hit the target with that initial value. You may be surprised at the outcome.

29. $x' = x - \sin(x)$, $t_1 = 5$, $x_1 = 2$.

30. $x' = x^2 - t$, $t_1 = 4$, $x_1 = 0$.

31. $x' = x(1 - x^2)$, $t_1 = 5$, $x_1 = 0.5$.

32. $x' = x \sin(x) + t$, $t_1 = 5$, $x_1 = 0$.

33. $x' = x \sin(x^2) + t$, $t_1 = 5$, $x_1 = 1$. In this case the authors were not able to hit the target. However, the exercise of trying is still worthwhile. We leave it to you to ponder why it is not possible.

Appendix: Downloading and Installing the Software Used in This Manual

The MATLAB programs `dfield`, `pplane`, and `odesolve`, and the solvers `eul`, `rk2`, and `rk4` described in this manual are not distributed with MATLAB. They are MATLAB function M-files and are available for download over the internet. There are versions of `dfield` and `pplane` written for use with all recent versions of MATLAB. However, `odesolve` is new, and only works with MATLAB ver 6.0 and later. The solvers are the same for all versions of MATLAB.

The following three step procedure will insure a correct installation, but the only important point is that the files must be saved as MATLAB M-files in a directory on the MATLAB path.

- Create a new directory with the name `odetools` (or choose a name of your own). It can be located anywhere on your directory tree, but put it somewhere where you can find it later.

- In your browser, go to `http://math.rice.edu/~dfield/`. For each file you wish to download, click on the link. In Internet Explorer, you are given the option to save the file. In Netscape, the file for the software will open in your browser, and you can save the file using the File menu. In either case, save the file with the subscript `.m` in the directory `odetools`.

- Open the path tool by executing the command `pathtool` in the MATLAB command window, or by selecting **File→ Set Path ...**. Follow the instructions for adding the directory `odetools` to the path. If you are asked if you want the change to be permanent, say yes.

From this point on, the programs will be available in MATLAB.

4. The Use of Function M-Files

In Chapter 2 we discussed script M-files, and how they can be used to facilitate the preparation of complicated graphics. Here we will introduce *function M-files*, which are very like script M-files, but can pass parameters and isolate variables. These two capabilities can be exploited to add new functions to MATLAB's already extensive list, and to extend our ability to perform computational experiments and prepare graphics.

In this chapter we will emphasize the use of MATLAB's built-in editor. If you use a different editor, and one of the authors does, then ignore these parts. However, the MATLAB editor has several connections with MATLAB iself that make it a good choice.

New Functions in MATLAB

It is very easy to add your own functions to the long list provided by MATLAB. Let's start with a very easy example.

Example 1. *Create a function M-file for the function* $f(x) = x^2 - 1$.

Execute `edit` at the command line to open the MATLAB editor[1] and enter the following three lines (one of them is blank, and is provided for readability):

```
function y = f(x)

y = x^2-1;
```

The first line of a function M-file must conform to the indicated format. It is this first line that distinguishes between a function M-file and a script M-file. The very first word must be the word `function`. The rest of the first line has the form

```
dependent_variables = function_name(independent_variables)
```

In the function `f`, `y = f(x)` indicates that x is the independent variable and y is the dependent variable.

The rest of the function M-file defines the function using the same syntax we have been using at the MATLAB prompt. Remember to put a semicolon at the end of lines in which computations are done. Otherwise the results will be printed in the Command Window.

When you save the file, you will be prompted to use the file name `f.m`. The file name should always be the same as the function name, followed by the suffix `.m`, so accept the suggestion. That's all there is to it. Now if you want to compute $f(3) = (3)^2 - 1 = 8$, simply enter `f(3)` at the MATLAB prompt.

```
>> f(3)
ans =
      8
```

---

[1] There are several ways to open the editor. Explore the Toolbar and the **Edit** menu.

Making Functions Array Smart. There is one important enhancement you will want to make to the M-file `f.m`. If you try to compute f on a matrix or a vector, you will find that it is not array smart.

```
>> x = 1:5
x =
     1    2    3    4    5
>> f(x)
??? Error using ==> ^
Matrix must be square.
```

This error message refers to the fact that we tried to compute x^2 for a vector x. We forgot to use array exponentiation instead of ordinary exponentiation. To edit the function to make it array smart we simply add one period:

```
function y = f(x)

y = x.^2-1;
```

Now the function can handle matrices. With the same vector x, we get

```
>> f(x)
ans =
     0    3    8   15   24
```

Notice that if x is a matrix, then so is x.^2. On the other hand, 1 is a number, so the difference x.^2-1 is not defined in ordinary matrix arithmetic. However, MATLAB allows it, and in cases like this will subtract 1 from every element of the matrix. This is a very useful feature in MATLAB.

Example 2. *An object thrown in the air obeys the initial value problem*

$$\frac{d^2y}{dt2} = -9.8 - \frac{dy}{dt}, \quad \text{with} \quad y(0) = 0 \quad \text{and} \quad y'(0) = 120.$$

When we solve this linear equation, we find that

$$y = -\frac{49}{5}t + \frac{649}{5}\left(1 - e^{-t}\right),$$

where y is the height in meters of the object above ground level after t seconds. Estimate the height of the ball after 5 seconds.

This is a perfect situation for a function M-file, particularly if you are interested in predicting the height of the ball at a number of additional times other than $t = 5$ seconds. Open a new M-file in the editor, enter

```
function y = height(t)
y = -(49/5)*t + (649/5)*(1 - exp(-t));
```

49

and save it as `height.m`. This time we needed no additional periods to make the function array smart. It is now a simple matter to find the height at $t = 5$ seconds.

```
>> y = height(5)
y =
    79.9254
```

Hence, the height of the ball at $t = 5$ seconds is 79.9254 meters.

Example 3. *Plot the height of the object in Example 2 versus time and use your graph to estimate the maximum height of the object and the time it takes the object to return to ground level.*

Although we could operate strictly from the command line to obtain a plot, we will will use a script M-file to do the work for us. Open a new M-file, enter the commands

```
close all
t = linspace(0,15,200);
y = height(t);
plot(t,y)
grid on
xlabel('time in seconds')
ylabel('height in meters')
title('Solution of y'''' = -9.8 - y'', y(0) = 0, y''(0) = 120')
```

and save it in a file named `height_drv.m`. Notice that, since the first line does not begin with the word `function`, this is a script M-file, not a function M-file.

The command `close all` will close all open figure windows. The purpose of this command is to prevent figure windows from accumulating as we execute versions of the script. However, you should be cautious about using it, since you may close figures you want to have open. Notice that we used the function `height` defined in Example 2 to calculate the y-values over the time interval $[0, 15]$.

You can execute the script M-file by typing `height_drv` at the MATLAB prompt. However, you can also select **Debug**→**Run**[2] from the editor menu. This menu item has the accelerator key F5. This means that after you edit the file, you can both save it and execute it with F5. The routine of editing, followed by F5 can be a great time saver as you refine an M-file.

Running your script M-file should produce an image similar to that shown in Figure 4.1. If your file contain errors, you might hear a "beep" or a "click" when you press the F5 button. An error message will be reported in the MATLAB command window, often as a link. Clicking this link will bring up the editor with the cursor on the line in the where the error occurs.

By examining the graph, we see that the object reaches a maximum height between 90 and 100 meters after about two or three seconds of flight. It returns to ground level (height zero) after approximately thirteen or fourteen seconds. You can use the zoom tool located on the toolbar in the figure window to obtain better estimates.

---

[2] This menu item changes to **Save and Run** if you have made changes since the last time you saved the file.

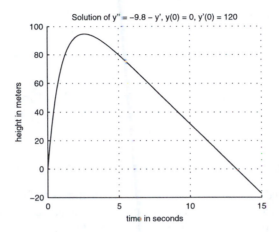

Figure 4.1. Plotting the solution of $y'' = -9.8 - y'$, $y(0) = 0$, $y'(0) = 120$.

Function M-files, Computational Exercises and Graphics

Some of you might wonder why we had to use *both* the function M-file height.m and the script M-file height_drv.m to produce Figure 4.1 in Example 2. It takes time to write files, and the accumulation of files clutters up the directory in which they are saved. Starting with version 6 of MATLAB there is a way around this problem.

Subfunctions in MATLAB 6. The answer is to define the function height as a *subfunction* with the file height_drv.m. However, subfunctions are not allowed in script M-files.

Open MATLAB's built-in editor, enter

```
function ch4examp3
close all
t = linspace(0,15,200);
y = height(t);
plot(t,y)
grid on
xlabel('time in seconds')
ylabel('height in meters')
title('Solution of y'''' = -9.8 - y'', y(0) = 0, y''(0) = 120')

function y = height(t)
y = -(49/5)*t + (649/5)*(1 - exp(-t));
```

and save it as ch4examp3.m. Use the F5 key to execute the file. This should produce a plot similar to that in Figure 4.1.

Notice that the code in ch4examp3 is precisely the same as the code in the two files height.m and height_drv.m created in Examples 2 and 3. However, this is a single function M-file with height included as a subfunction.

51

Functions of functions — MATLAB's `funfun` directory. MATLAB has a large number of functions that act on functions. If you type

```
>> help funfun
```

at MATLAB's prompt, you will see a list of MATLAB routines for finding zeros and extrema of functions, tools for numerical integration, and many others. There is a suite of routines for solving differential equations, which we will look at in Chapter 8. For help on any of these functions, say `fzero`, execute `help fzero` and learn how to find the zeros of a function.

Example 4. *Use `fzero` to find the time it takes the object in Example 2 to return to ground level.*

The help command indicates that "X = FZERO(FUN,X0) tries to find a zero of the function FUN near X0." Thus, `fzero` needs two inputs, the name of a function, and an approximate value of the zero. We will use the function `height` from Example 2, and from Figure 4.1 we estimate that `height` has a zero near $t = 13$. In all versions of MATLAB you can invoke `fzero` with the command

```
>> t = fzero('height',13)
t =
    13.2449
```

Notice that the function name must be entered between single quotes. Starting with version 6 of MATLAB, the use of

```
>> t = fzero(@height,13)
t =
    13.2449
```

is encouraged. The expression `@height` is a *function handle* for the function `height`. In either case, we find that the object returns to ground level after about 13.2 seconds. This certainly agrees with that we see in Figure 4.1.

Example 5. *Find the maximum height reached by the object in Example 2.*

In the list provided by `help funfun` we find that MATLAB doesn't provide a function for finding the maximum of a function. However, the routine `fminbnd` will find a local minimum, and a local maximum of a function can be determined by finding a local minimum of the negative of the function.

Using `help fminbnd`, we learn that finding the minimum of a function on a given interval requires the input of a function and the beginning and endpoints of the interval in which to find the minimum. Consequently, we will apply `fminbnd` to the negative of `height` on the interval $[0, 5]$. One approach would be to write a new function M-file encoding the negative of $-(49/5)t + (649/5)(1 - e^{-t})$. However, let's pursue an alternative.

MATLAB provides a perfect solution for the situation when you need a function, but do not want to go to the trouble of writing a function M-file, called an *inline function*.[3] We could create our inline

---

[3] Type `help inline` to get a thorough explanation of the use of inline functions.

function for the negative of `height` as

```
>> f = inline('(49/5)*t - (649/5)*(1 - exp(-t))','t')
f =
     Inline function:
     f(t) = (49/5)*t - (649/5)*(1 - exp(-t))
```

However, since we have already created `height`, it is easier to use the inline function

```
>> f = inline('-height(t)','t')
f =
     Inline function:
     f(t) = -height(t)
```

From Figure 4.1 we estimate that the maximum occurs between $t = 0$ and $t = 5$, so the command

```
>> t = fminbnd(f,0,5)
t =
     2.5836
```

finds the time at which the maximum occurs. Notice that there are no single quotes around `f` in `fminbnd(f,1,5)`, so inline functions are treated differently from function M-files as inputs. Alternatively, the syntax `t = fminbnd(inline('-height(t)'),1,5)` will do the job in a single command.

The maximum height is now easily calculated by

```
>> h = -f(t)
h =
     94.6806
```

or by `h = height(t)`. Compare these results with Figure 4.1.

Using function M-files to pass parameters. The ability of function M-files to have input variables can be used to advantage when producing graphics. We will present a simple example.

Example 6. *Write a function M-file that will plot the solution to the initial value problem*

$$y' + y = 2t \cos t \quad \text{with} \quad y(0) = y_0$$

over the interval $[0, 4\pi]$. *Write the file with* y_0 *as an input variable.*

The equation is linear, and we find that the general solution is $y(t) = t(\cos t + \sin t) - \sin t + Ce^{-t}$. At $t = 0$, we find that $y_0 = y(0) = C$. Thus the function M-file

```
function ch4examp6(y0)
t = linspace(0,4*pi);
y = t.*(cos(t) + sin(t)) - sin(t) + y0*exp(-t);
plot(t,y), shg
```

53

will produce the required graph when `ch4examp6(y0)` is executed at the command line. Of course, you will want to add formatting commands to your function M-file. The result is shown in Figure 4.2. The one tricky formatting command is the one that produces the title. The command

```
title(['y'' + y = 2tcos(t)    with y(0) = ', num2str(y0), '.'])
```

requires the concatenation of text strings, as explained in the section "Text Strings in MATLAB" in Chapter 2.

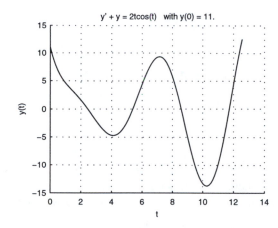

Figure 4.2. The output of `ch4examp6(11)`.

Functions of Several Variables

MATLAB allows multiple input and output variables in function M-files.

Example 7. *Write a function M-file for the function defined by $f(t, x) = x^2 - t$.*

We need to choose a name for this function, so let's use the mnemonic xsqmt ("x squared minus t"). Then the following M-file describes this function:

```
function y = xsqmt(t,x)
% This is Example 7 of Chapter 4.
y=x.^2-t;
```

The first line of the file starts with the word `function`. After that comes y=xsqmt(t,x), meaning that y is the dependent variable, and both t and x are independent variables.

Everything on a line in an M-file after a percentage sign (%) is ignored by MATLAB. This can be utilized to put comments in a file, as we have done in the function xsqmt. Comment lines are useful ways to document your M-files. It is amazing how quickly we forget why we did things the way we did. Comment lines immediately after the function statement can be read using the MATLAB `help` command.

For example, if `xsqmt` is defined as above, saved as `xsqmt.m`, then `help xsqmt` gives the following response.

```
>> help xsqmt

  This is Example 7 of Chapter 4.
```

You can evaluate the function at $(t, x) = (2, 5)$ as follows.

```
>> xsqmt(2,5)
ans =
    23
```

Naming and Organizing Files

Once you learn how to write M-files, they tend to accumulate rapidly. It is a good idea to give them appropriate names and to organize them into a directory structure that will make them easy to find. Of course you will choose a method that suits you. We will present some general considerations.

As one starts to use files, it is typical to save them all to the same directory. That is fine if you only have a few. When the number gets large enough that you have trouble identifying them, it becomes necessary to create new directories, and it is a good idea to use the directory system to differentiate the files. For example, you might create a directory named `manual` in which you save all files related to this Manual.

The method of creating directories or folders depends on your operating system. You can use the MATLAB command `mkdir`, but this is usually less convenient. Once you have more than one directory, you will need to know how to navigate through them. For one thing, MATLAB has a "current directory" in which it operates. This directory is displayed in the Current Directory popup menu on the toolbar of the command window. You can also discover the current directory by executing `pwd`. The popup menu lists all of the directories you have visited recently, making it easy to navigate among them. You can change to a directory not on the popup menu using the browse button, labeled with an ellipsis (. . .). You can also use the command `cd`.

The MATLAB path. After you have created a couple of directories, you will have to become familiar with the MATLAB path. Difficulties will arise in one of two ways. The first happens when you try to execute an M-file and get the error message

```
>> phony
??? Undefined function or variable 'phony'.
```

However, you are sure you created `phony.m` just yesterday, and it worked fine. The problem here is probably that `phony.m` is in a directory that is not on the MATLAB path. The second difficulty that can arise is that you execute an M-file, and, while it works, it is not doing what you know it should. In this case you probably have two files named `phony.m`, in different directories, and MATLAB is finding the wrong one.

The MATLAB path is the list of directories that MATLAB searches when it is looking for an M-file, in the order that they are searched. You can find the path using the pathtool,[4] which can be accessed using the command `pathtool` or by choosing the menu item **File→Set Path...** . The first directory on the path is always the current directory, so you can be sure you are executing the correct M-file if you find out what directory it is in, and then change directories so that it is the current directory. It is also possible to add directories to the MATLAB path using the pathtool. Files that you use frequently should be in directories that are permanently on the path.

If you find that you have two files with the same name, say phony, you can find out which one is higher on the path with the command `which phony`. MATLAB will respond with the complete address of the first instance of `phony.m` on the MATLAB path. If the address is not the one you expect, you know that you have a name conflict. You could delete the offending file (probably not a good idea), you could change the name of one of the files, or you could just change directories so that the directory containing the right version of `phony.m` is the current directory.

Either of the commands `ls` or `dir` will display a list of the files in the current directory. You can display the contents of the M-file `phony.m` on the command window with the command `type phony`. This works very well with short files, but not too well with long ones.

Naming Functions and Variables. All of us typically refer to functions with one letter names, such as f or g. However, many important and commonly used functions have longer names. Think of the trigonometric functions, the logarithm, and the hyperbolic functions as examples. It is a good idea to follow that practice when naming your MATLAB M-files. Giving your function M-files distinctive names helps to avoid name conflicts and might help you remember their functionality at a later date. For example, in Example 3 the name ch4examp3 was carefully chosen to reflect the fact that the file is related to Example 3 in Chapter 4 of this Manual.

The name of a function M-file has the form `function_name.m`, where `function_name` is the name you choose to call the function. While this name can be almost anything, there are a few rules.

- The name must start with a letter (either upper case or lower case).

- The name must consist entirely of letters, numerals, and underscores (_). No other symbols are allowed. In particular, no periods are allowed.

- The name can be arbitrarily long, but MATLAB will only remember the first 31 characters.

- Do not use names already in use such as `cos`, `plot`, or `dfield6`. If you do, MATLAB will not complain, but you will eventually suffer from the name duplication problem described earlier.

The variable names used in a function M-file must satisfy the same rules that apply to the names of M-files. Other than that they are arbitrary. As a result the file `f.m` of Example 1 could have been written as

```
function stink = funn(skunk)
stink = skunk.^2 - 1;
```

Save this file as `funn.m` and with `x = 1:5` execute

---

[4] Or you can type `path` at the MATLAB prompt.

```
>> funn(x)
ans =
     0    3    8   15   24
```

Compare this to the output produced earlier with f(x) and note that there is no difference. It is important to realize that the variable names used in function M-files are *local* to these files; i.e., they are not recognized outside of the M-files themselves. See Exercises 12 and 13. In addition variables that have been defined at the command line are not available within function M-files. Script M-files, on the other hand, do not isolate variables.

Exercises

Find the solution of each of the initial value problems in Exercises 1–4, then craft an inline function for the solution. After insuring that your inline function is "array smart," use the inline function to (i) plot the solution of the initial value problem on the given interval, and (ii) evaluate the solution at the right endpoint of the given interval.

1. $y' = -t/y$, with $y(0) = 4$, on $[0, 4]$.
2. $y' = -2ty^2$, with $y(0) = 1$, on $[0, 5]$.
3. $y' + y = \cos t$, with $y(0) = 1$, on $[0, 5\pi]$.
4. $y' + 2y = \sin(2t)$, with $y(0) = 1$, on $[0, 4\pi]$.

Find the solution of each of the initial value problems in Exercises 5–8, then craft a function M-file for the solution. After insuring that your function M-file is "array smart," use the function to (i) plot the solution of the initial value problem on the given interval, and (ii) evaluate the solution at the right endpoint of the given interval.

5. $y' = (1 - t)y$, with $y(0) = 1$, on $[0, 4]$.
6. $y' = y \sin(2t)$, with $y(0) = 1$, on $[0, 2\pi]$.
7. $y' = y/t + t$, $y(1) = -1$, on $[0, 3]$.
8. $y' + t^2 y = 3$, with $y(0) = -2$, on $[0, 5]$.

Find the solution of the given initial value problem in Exercises 9–12, then craft a function M-file for the solution. After insuring that your function M-file is "array smart," plot the solution on the given interval. Use MATLAB's fminbnd function to find the minimum value of the solution on the given interval and the time at which it occurs. Use the figure window's "Insert Text" and "Insert Arrow" tools to annotate your plot with these findings.[5]

9. $y' + 2ty = -e^{-t^2}$, with $y(0) = 0$, on $[0, 2]$.
10. $y' - y/t = -t \sin t$, with $y(\pi) = -\pi$, on $[0, 2\pi]$.
11. $y' = 1 + 2y/t$, with $y(1) = -3/4$, on $[1, 4]$.
12. $y' + y = t^2$, with $y(0) = 1$, on $[0, 2]$.

Find the solution of the given initial value problem in Exercises 13–16, then craft a function M-file for the solution. After insuring that your function M-file is "array smart," plot the solution on the given interval. Use MATLAB's fminbnd function and the technique of Example 5 to find the maximum value of the solution on the given interval and the time at which it occurs. Use the figure window's "Insert Text" and "Insert Arrow" tools to annotate your plot with these findings.

13. $ty' - y = t^2 \cos(t)$, with $y(\pi) = 0$, on $[0, 2\pi]$.
14. $y' = y \cos(t)$, with $y(0) = 1$, on $[0, 2\pi]$.
15. $y' = (1 + y) \sin(t)$, with $y(\pi) = 0$, on $[0, 2\pi]$.

---

[5] If the figure toolbar is not visible, select **View→ Figure Toolbar** from the figure menu.

16. $y' + t^2 y = 3$, with $y(0) = 1$, on $[0, 3]$.

17. The differential equation $y' + y = \cos(4t)$ has the general solution $y = (1/17)\cos(4t) + (4/17)\sin(4t) + Ce^{-t}$. Craft a function M-file for this solution as follows.

```
function y = f(t,C)
y = 1/17*cos(4*t) + 4/17*sin(4*t) + C*exp(-t);
```

Plot a *family of solutions* for the differential equation with the following script.

```
t = linspace(0,2*pi,500);
y = [];
for C = -20:5:20
    y = [y;f(t,C)];
end
plot(t,y)
```

Follow the lead of Exercise 17 to plot the given families of solutions in Exercises 18–23. Use the indicated constants and extend your plot over the given time interval.

18. $y = 2/(t^2 + C)$, $C = 1, 2, 3, 4, 5$, on $[-2, 2]$.

19. $y = (1/4)t^2 - (1/8)t + 1/32 + Ce^{-4t}$, $C = 1, 2, 3, 4, 5$, on $[0, 2]$.

20. $y = Cte^{-t^2}$, $C = -3, -2, -1, 0, 1, 2, 3$, on $[-2, 2]$.

21. $y = (t/2)(\cos(t) + \sin(t)) - (1/2)\sin(t) + Ce^{-t}$, $C = -3, -2, -1, 0, 1, 2, 3$, on $[0, 10]$.

22. $y = -1 + \cos(t) + (C + t/2)\sin(t)$, $C = 0, 1, 2, 3, 4, 5$, on $[0, 6\pi]$.

23. $y = (1/10)\sin(t) + (1/5)\cos(t) + e^{-t}(C\sin(2t) - (1/5)\cos(2t))$, $C = -3, -2, -1, 0, 1, 2, 3$, on $[0, 2\pi]$.

24. A tank initially contains 20 gallons of pure water. A salt solution containing 2 lb/gal flows into the tank at a rate of 3 gal/min. A drain is open at the bottom of the tank through which flows salt solution from the tank at a rate of 2 gal/min. Set up and solve an initial value problem modeling the amount of salt in the tank at time t minutes. Write an inline function for your solution and use it to find the salt content in the tank at $t = 30$ minutes.

25. Tank A initially contains 40 gallons of a salt solution. Pure water is poured into tank A at a rate of 2 gal/min. A drain is opened at the bottom of tank A so that salt solution flows directly from tank A into tank B at 2 gal/min, keeping the volume of the salt solution in tank A constant over time. Tank B initially contains 100 gallons of pure water. A drain is opened at the bottom of tank B so that the volume of solution in tank B remains constant over time. If x and y represent the salt content in tanks A and B, respectively, show that

$$\frac{dx}{dt} = -\frac{x}{25} \quad \text{and} \quad \frac{dy}{dt} = \frac{x}{25} - \frac{y}{50}.$$

Given that $x(0) = 20$, solve the first equation for x, substitute the result into the second equation, then show that

$$y = 40e^{-t/50} - 40e^{-t/25}.$$

Write a function M-file to evaluate y at time t. Following the technique used in Examples 3-5, use your function to (i) construct a plot of the salt content in tank B over the time interval $[0, 100]$, and (ii) use `fminbnd` to find the maximum salt content in tank B and the time at which this occurs.

26. An object thrown into the air is known to obey the initial value problem

$$y'' = -9.8 - 0.05y', \qquad y(0) = 0, \qquad y'(0) = 200,$$

where y is the height of the ball above ground (in meters) at t seconds. The solution of this initial value problem is

$$y(t) = -196t + 7920(1 - e^{-t/20}).$$

58

Following the lead of Examples 2-5 in the text, plot this solution, then find its maximum height and the time at which it occurs. Then find the time it takes the object to return to the ground. Question: Does the time it takes the object to reach its maximum height equal the time it takes the object to return to ground level from this maximum height? In other words, does it take the object the same amount of time to go up as it takes the object to come down?

27. A simple RC-circuit with emf $V(t) = 3\cos(\omega t)$ is modeled by the initial value problem

$$RCV_C' + V_C = 3\cos(\omega t), \qquad V_C(0) = 0,$$

where R is the resistance, C the capacitance, ω is the driving frequency, and V_C is the voltage response across the capacitor. Show that the solution is

$$V_C(t) = \frac{RC\omega \sin \omega t + \cos \omega t - e^{-t/(RC)}}{1 + R^2 C^2 \omega^2}.$$

Create a function M-file for this solution as follows.

```
function V = f(t,R,C,w)
V = (R*C*w*sin(w*t) + cos(w*t) - exp(-t/(R*C)))/(1 + R^2*C^2*w^2);
```

Now, call this function with the following script.

```
R = 1.2; C = 1; w = 1;
t = linspace(0,2*pi,1000);
V = f(t,R,C,w);
plot(t,V)
```

Run this script for $\omega = 1, 2, 4$, and 8, keeping $R = 1.2$ ohms and $C = 1$ farad constant. What happens to the amplitude of the voltage response across the capacitor as the frequency of the emf is increased? Why do you think this circuit is called a *low pass filter?*

28. Write a function M-file to calculate $f(z) = \sqrt{|z|}$. Use it to graph f over the interval $(-3, 2)$.

29. Write a function M-file to compute the function $f(t) = t^{1/3}$. Use it to graph the function over the interval $[-1, 1]$. (This is not as easy as it looks. Read the section on complex arithmetic in Chapter 1.)

30. Write a function M-file to calculate the function $f(t) = e^{-t}(t^2 + 4e^t - 5)$.

    a) Use the M-file to graph f over the interval $(-2.5, 3)$.

    b) Find all of the zeros in the interval $(-2.5, 3)$ of the function f defined in the previous exercise.

31. For the function funn defined in this chapter find all solutions to the equation funn$(x) = 5$ in the interval $[-5, 5]$.

32. Write an array smart function M-file to compute $f(t) = 5$. **Remark:** You will find this exercise more difficult than it looks. There are a variety of tricks that will work, but the methods that are most consistent with the ways used for non constant functions (and therefore are most useful in programming applications) use the MATLAB commands size and ones. Use help on these commands.

33. Variables defined in functions are *local* to the function. To get a feel for what this means, create the function M-file

```
function y = fcn(x)
A = 2;
y = A^x;
```

Save the file as fcn.m. In MATLAB's command window, enter the following commands.

```
A = 3; x = 5;
y = fcn(x)
A^x
```

Explain the discrepancy between the last two outputs. What is the current value of A in MATLAB's workspace?

34. You can make variables *global*, allowing functions access to the variables in MATLAB's workspace. To get a feel for what this means, create the function M-file

```
function y = gcn(x)
global A
A = 2;
y = A^x;
```

Save the file as gcn.m. In MATLAB's command window, enter the following commands.

```
global A
A = 3; x = 5;
y = gcn(x)
A^x
```

Why are these last two outputs identical? What is the current value of A in MATLAB's workspace?

5.   Numerical Methods for ODEs

Numerical methods for solving ordinary differential equations are discussed in many textbooks. Here we will discuss how to use some of them in MATLAB. In particular, we will examine how a reduction in the "step size" used by a particular algorithm reduces the error of the numerical solution, but only at a cost of increased computation time. In line with the philosophy that we are not emphasizing programming in this manual, MATLAB routines for these numerical methods are made available.

Our discussions of numerical methods will be very brief and incomplete. The assumption is that the reader is using a textbook in which these methods are described in more detail.

Euler's Method

We want to find an approximate solution to the initial value problem $y' = f(t, y)$, with $y(a) = y_0$, on the interval $[a, b]$. In Euler's very geometric method, we go along the tangent line to the graph of the solution to find the next approximation. We start by setting $t_0 = a$ and choosing a step size h. Then we inductively define

$$
\begin{aligned}
y_{k+1} &= y_k + hf(t_k, y_k), \\
t_{k+1} &= t_k + h,
\end{aligned}
\tag{5.1}
$$

for $k = 0, 1, 2, \ldots, N$, where N is large enough so that $t_N \geq b$. This algorithm is available in the MATLAB command eul[1].

Example 1. *Use Euler's method to plot the solution of the initial value problem*

$$
y' = y + t, \quad y(0) = 1,
\tag{5.2}
$$

on the interval $[0, 3]$.

Note that the differential equation in (5.2) is in normal form $y' = f(t, y)$, where $f(t, y) = y + t$. Before invoking the eul routine, you must first write a function M-file

```
function yprime = yplust(t,y)
yprime = y + t;
```

to compute the right-hand side $f(t, y) = y + t$. Save the function M-file with the name yplust.m. Before continuing, it is always wise to test that your function M-file is returning the proper output. For example, $f(1, 2) = 2 + 1$, or 3, and

```
>> yplust(1,2)
   ans = 3
```

---

[1] The function eul, as well as rk2 and rk4 described later in this chapter, are function M-files which are not distributed with MATLAB. For example, type help eul to see if eul is installed correctly on your computer. If not, see the Appendix to Chapter 3 for instructions on how to obtain them. The M-files defining these commands are printed in the appendix to this chapter for the illumination of the reader.

If you don't receive this output, check your function `yplust` for coding errors. If all looks well, and you still aren't getting the proper output, refer to the section headed **The MATLAB Path** in Chapter 4.

Now that your function M-file is operational, it's time to invoke the `eul` routine. The general syntax is

```
[t,y]=eul(@yplust,tspan,y0,stepsize)
```

where `yplust` is the name of the function M-file, `tspan` is the vector `[t0,tfinal]` containing the initial and final time conditions, `y0` is the y-value of the initial condition, and `stepsize` is the step size to be used in Euler's method.[2] Actually, the step size is an optional parameter. If you enter

```
[t,y]=eul(@yplust,tspan,y0)
```

the program will choose a default step size equal to `(tfinal-t0)/100`.

It is also possible to invoke `eul` with the command `[t,y]=eul('yplust',tspan,y0,stepsize)`. The only difference is the use of `'yplust'` instead of `@yplust`. If you are using a version of MATLAB prior to version 6, you have no choice. You must use `'yplust'`. With version 6 you are given the choice. Using the *function handle* `@yplust` will result in speedier solution. This speedup is not at all important for the simple equations we solve in this chapter. However, it becomes very important when you are solving more complicated systems of ODEs. The use of function handles also provides us with more flexibility of use, as we will explain in Chapter 7. We will systematically use the function handle notation, leaving it to you to make the replacement, if necessary.

The following commands will produce an approximate solution of the initial value problem (5.2) similar to that shown in Figure 5.1.

```
>> [t,y] = eul(@yplust,[0,3],1,1);
>> plot(t,y,'.-')
```

We can arrange the output in two columns with the command

```
>> [t,y]
ans =
        0     1
        1     2
        2     5
        3    12
```

Notice that the t_k's given in the first column differ by 1, the stepsize. The y_k's computed inductively by the algorithm in (5.1) are in the second column. In Figure 5.1 these four points are plotted and connected with straight lines.

---

[2] You can use an inline function defined with the command `f = inline('y+t','t','y');` instead of the function M-file. Then the syntax for using `eul` is `[t,y] = eul(f,tspan,y0,stepsize)`.

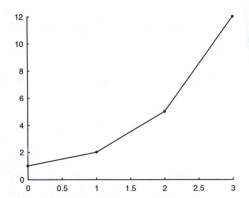

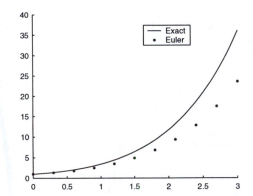

Figure 5.1. Euler's solution of initial value problem (5.1).

Figure 5.2. Euler versus exact solution (Step size $h = 0.3$).

Euler's Method Versus the Exact Solution

In this section we want to examine visually the error in the numerical solution obtained from Euler's method.

Example 2. *Re-examine the initial value problem*

$$y' = y + t, \quad y(0) = 1. \tag{5.3}$$

Compare the exact solution and the numerical solution obtained from Euler's method on the interval $[0, 3]$. *Use a step size of* $h = 0.3$.

The initial value problem in (5.3) is linear and easily solved:

$$y = -t - 1 + 2e^t. \tag{5.4}$$

The Euler's method solution of the initial value problem in (5.3) with step size 0.3 is found by

```
>> h = 0.3;
>> [teuler,yeuler] = eul(@yplust,[0,3],1,h);
```

We compute the exact solution using equation (5.4), but we want to choose a finer time increment on the interval $[0, 3]$ so that the exact solution will have the appearance of a smooth curve.

```
>> t = 0:.05:3;
>> y = -t - 1 + 2*exp(t);
```

The command

```
>> plot(t,y,teuler,yeuler,'.')
>> legend('Exact','Euler')
```

produces Figure 5.2, a visual image of the accuracy of Euler's method using the step size $h = 0.3$ The accuracy is not too good in this case.

Changing the Step Size — Using Script M-Files

We would like to repeat the process in Example 2 with different step sizes to analyze graphically how step size affects the accuracy of the approximation.

Example 3. *Examine the numerical solutions provided by Euler's method for the initial value problem*

$$y' = y + t, \quad y(0) = 1,$$

on the interval $[0, 3]$. *Use step sizes* $h = 0.2, 0.1, 0.05,$ *and* $0.025.$

It will quickly get tedious to type the required commands into the Command Window. This is a good place to use a script M-file, as discussed in Chapter 2. We collect all of the commands and type them once into an editor. To be precise, this file should contain the lines

```
[teuler,yeuler] = eul(@yplust,[0,3],1,h);
t = 0:.05:3;
y = -t - 1 + 2*exp(t);
plot(t,y,teuler,yeuler,'.')
legend('Exact','Euler')
shg
```

We save the file as `batch1.m` (or whatever else we want to call it as long as it meets the requirements for function names). Then executing `>> h = 0.2; batch1` produces Figure 5.3, while the command `>> h = 0.05; batch1` produces Figure 5.5.

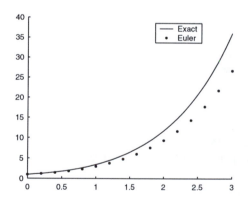

Figure 5.3. Euler versus exact solution (Step size $h = 0.2$).

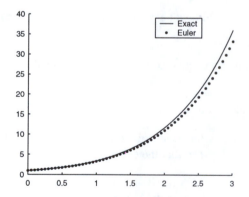

Figure 5.4. Euler versus exact solution (Step size $h = 0.05$).

Repeating this command with various values of the step size h allows us to visually examine the effect of choosing smaller and smaller step sizes. Note how the error decreases, but the computation time increases as you reduce the step size.

64

Numerical Error Analysis

In this section we analyze the error inherent in Euler's method numerically as well as graphically by computing the *absolute error* at each step, defined as the absolute value of the difference between the actual solution and the numerical solution at that step. We will continue to look at the error graphically.

Example 4. *Consider again the initial value problem*

$$y' = y + t, \quad y(0) = 1,$$

on the interval $[0, 3]$. *Record the maximum error in the Euler's solution for the step sizes* $h = 0.2, 0.1, 0.05,$ *and* 0.025.

We record the error at each step by adjusting the script file `batch1.m` from Example 3. We need to evaluate the exact solution at the same points at which we have the approximate values calculated, i.e. at the points in the vector `teuler`. This is easy. Simply enter `z = -teuler - 1 + 2*exp(teuler)`. Then to compare with the approximate values, we look at the difference of the two vectors `z` and `yeuler`. We are only interested in the magnitude of the error and not the sign. The command `abs(z-yeuler)` yields a vector containing the absolute value of the error made at each step of the Euler computation. Finally, we compute the largest of these errors with the MATLAB command `maxerror = max(abs(z-yeuler))`. We enter this command without an ending semi-colon, because we want to see the result in the command window. Therefore, to effect these changes we alter the file `batch1.m` to

```
[teuler,yeuler] = eul(@yplust,[0,3],1,h);
t = 0:.05:3;
y = -t - 1 + 2*exp(t);
plot(t,y,teuler,yeuler,'.')
legend('Exact','Euler')
shg
z = -teuler - 1 + 2*exp(teuler);
maxerror = max(abs(z - yeuler))
```

Save this file as `batch2.m`. A typical session might show the following in the command window:

```
>> h = 0.2; batch2
maxerror =
     9.3570
>> h = 0.1; batch2
maxerror =
     5.2723
>> h = 0.05; batch2
maxerror =
     2.8127
```

Of course, each of these commands will result in a visual display of the error in the figure window as well. Thus, we can very easily examine the error in Euler's method both graphically and quantitatively.

If you have executed all of the commands up to now, you will have noticed that decreasing the step size has the effect of reducing the error. But at what expense? If you continue to execute `batch2`, halving the step size each time, you will notice that the routine takes more and more time to complete. In fact, halving the step size doubles the number of computations that must take place and therefore doubles the time needed. Therefore, decreasing the step size used in Euler's method is expensive, in terms of computer time. However, notice that halving the step size results in a maximum error almost exactly half the previous one.

At this point it would be illuminating to see a plot of how the maximum error changes versus the step size.

Example 5. *Consider again the initial value problem*

$$y' = y + t, \quad y(0) = 1.$$

Sketch a graph of the maximum error in Euler's method versus the step size.

To do this using MATLAB, we have to devise a way to capture the various step sizes and the associated errors into vectors and then use an appropriate plotting command. We will call the vector of step sizes h_vect and the vector of errors err_vect. The following modification of batch2.m will do the job. Save the file as batch3.m

```
h_vect = []; err_vect = []; h = 1;
t = 0:.05:3;
y = -t - 1 + 2*exp(t);
for k = 1:8
    [teuler,yeuler] = eul(@yplust,[0,3],1,h);
    plot(t,y,teuler,yeuler,'.')
    legend('Exact','Euler',2)
    shg
    z = -teuler - 1 + 2*exp(teuler);
    maxerror = max(abs(z - yeuler));
    h_vect = [h_vect,h]
    err_vect = [err_vect,maxerror]
    pause
    h = h/2;
end
```

The first line initializes h_vect, err_vect, and the starting step size h = 1. Since at the beginning they contain no information, h_vect and err_vect are defined to be empty matrices, as indicated by []. The next two lines compute the exact solution for later plotting. The main work is done in the for loop. This comprises all of the indented steps between for k = 1:8 and end. The first four lines compute the approximate solution, and plot it in comparison with the exact solution. Notice the third input to the legend command. This results in the legend being placed in the upper left corner, where it does not interfere with the graph. The maximum error is computed next and the new step size and error are added to the vectors h_vect and err_vect. The command pause stops the execution and gives you time to observe the figure and the output in the command window. When you are finished examining the figure and output, hit any key (the space bar will do) to advance to the next step. The final line in the for loop halves the step size in preparation for the next step. Notice that the for loop is executed 8 times with different values of k

Now we are ready to go. Executing `batch3` results in a visual display similar to Figures 5.3 or 5.4 at each step. There is also output on the command window because we have not ended the `h_vect = ...` and `err_vect = ...` commands with semi-colons. By hitting a key twice, we get three steps of the `for` loop, and the output

```
>> batch3
h_vect =
     1
err_vect =
   24.1711
h_vect =
     1.0000     0.5000
err_vect =
   24.1711    17.3898
h_vect =
     1.0000     0.5000     0.2500
err_vect =
   24.1711    17.3898    11.0672
```

at the command line. Continuing to step through `batch3`, we get more and more data. After the seventh iteration the output is

```
h_vect =
     1.0000     0.5000     0.2500     0.1250     0.0625     0.0313     0.0156
err_vect =
   24.1711    17.3898    11.0672     6.3887     3.4581     1.8030     0.9211
```

Notice that the maximum error is approximately halved on each step, at least on the last few steps.

After all eight steps we are ready to plot the error data. This can be done with any of the commands `plot`, `semilogx`, `semilogy`, or `loglog`. Try them all to see which you like best. However, since the maximum error seems to be halved each time the step size is halved, we expect a power law relationship. Therefore we will use a loglog graph. The commands

```
>> loglog(h_vect,err_vect)
>> xlabel('Step size')
>> ylabel('Maximum error')
>> title('Maximum error vs. step size for Euler''s method')
>> grid
>> axis tight
```

produce Figure 5.5. The command `axis tight` sets the axis limits to the range of the data.

Notice that the graph of the logarithm of the maximum error versus the logarithm of the step size is approximately linear. If the graph were truly linear, we would have a linear relationship between the logarithms of the coordinates, like

$$\log_{10}(\text{maximum\_error}) = A + B \log_{10}(\text{step\_size}),$$

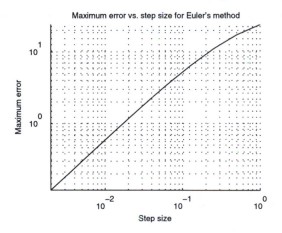

Figure 5.5. Analysis of the error in Euler's method.

for some constants A and B.[3]

If we exponentiate this relationship, we get

$$\text{maximum\_error} = 10^A \cdot (\text{step\_size})^B = C(\text{step\_size})^B,$$

where $C = 10^A$. Thus a linear relationship with slope B on a loglog graph means that there is a power law relationship between the coordinates with the exponent being B. Indeed, we will see in Exercise 21 that Euler's method is a first order method. This means that if y_k is the calculated approximation at t_k, and $y(t_k)$ is the exact solution evaluated at t_k, then $|y(t_k) - y_k|$ is the error being made, and it satisfies an inequality of the form

$$|y(t_k) - y_k| \le C|h|, \quad \text{for all } k, \tag{5.5}$$

where C is a constant. This inequality satisfied by the error is reflected in the nearly linear relationship we see in Figure 5.5, and it is the reason why we chose a loglog graph.

Hopefully, you now understand two key points about Euler's method. You can increase the accuracy of the solution by decreasing the step size, but you pay for the increase in accuracy with an increase in the number of computational steps and therefore in the computing time.

The Second Order Runge-Kutta Method

What is needed are solution routines that will provide more accuracy without having to drastically reduce the step size and therefore increase the computation time. The Runge-Kutta routines are designed to do just that. The simplest of these, the Runge-Kutta method of order two, is sometimes called the improved Euler's method. Again, we want to find a approximate solution of the initial value problem

---

[3] We are using the logarithm to base 10 instead of the natural logarithm because it is the basis of the loglog graph. However there is very little difference in this case. Since $\ln x = \log_{10} x \cdot \ln 10$, the inequality is true with $\log_{10}$ replaced by $\ln$, with the same slope B and A replaced by $A \cdot \ln 10$.

$y' = f(t, y)$, with $y(a) = y_0$, on the interval $[a, b]$. As before, we set $t_0 = a$, choose a step size h, and then inductively define

$$s_1 = f(t_k, y_k),$$
$$s_2 = f(t_k + h, y_k + hs_1),$$
$$y_{k+1} = y_k + h(s_1 + s_2)/2,$$
$$t_{k+1} = t_k + h,$$

for $k = 0, 1, 2, \ldots, N$, where N is large enough so that $t_N \geq b$. This algorithm is available in the M-file `rk2.m`, which is listed in the appendix of this chapter.

The syntax for using `rk2` is exactly the same as the syntax used for `eul`. As the name indicates, it is a second order method, so, if y_k is the calculated approximation at t_k, and $y(t_k)$ is the exact solution evaluated at t_k, then there is a constant C such that

$$|y(t_k) - y_k| \leq C|h|^2, \quad \text{for all } k. \tag{5.6}$$

Notice that the error is bounded by a constant times the square of the step size. As we did for Euler's method, we have a power inequality, but now the power is 2 instead of 1. We should expect a nearly linear relationship between the maximum error and the step size in a loglog graph, but with a steeper slope than we observed for Euler's method.

The error in the second order Runge-Kutta method can be examined experimentally using the same techniques that we illustrated for Euler's method. Modify the script files, `batch1.m`, `batch2.m`, and `batch3.m`, replacing the command `eul` with `rk2` everywhere it occurs, and changing all references to euler to refer to rk2 instead. You might also want to change the names of the files to indicate that they now refer to `rk2`. Replay the commands in Examples 3, 4, and 5 using your newly modified routines.

The Fourth Order Runge-Kutta Method

This is the final method we want to consider. We want to find an approximate solution to the initial value problem $y' = f(t, y)$, with $y(a) = y_0$, on the interval $[a, b]$. We set $t_0 = a$ and choose a step size h. Then we inductively define

$$s_1 = f(t_k, y_k),$$
$$s_2 = f(t_k + h/2, y_k + hs_1/2),$$
$$s_3 = f(t_k + h/2, y_k + hs_2/2),$$
$$s_4 = f(t_k + h, y_k + hs_3),$$
$$y_{k+1} = y_k + h(s_1 + 2s_2 + 2s_3 + s_4)/6,$$
$$t_{k+1} = t_k + h,$$

for $k = 0, 1, 2, \ldots, N$, where N is large enough so that $t_N \geq b$. This algorithm is available in the M-file `rk4.m`, which is listed in the appendix of this chapter.

The syntax for using `rk4` is exactly the same as that required by `eul` or `rk2`. As the name indicates, it is a fourth order method, so, if y_k is the calculated approximation at t_k, and $y(t_k)$ is the exact solution evaluated at t_k, there is a constant C such that

$$|y(t_k) - y_k| \leq C|h|^4, \quad \text{for all } k. \tag{5.7}$$

The error in the fourth order Runge-Kutta method can be examined experimentally using the same techniques that we illustrated for Euler's and the Runge-Kutta 2 methods. Again, you can easily modify the script files, batch1.m, batch2.m, and batch3.m to create files to evaluate the errors in the Runge-Kutta 4 algorithm. Since the error is bounded by a constant times the fourth power of the step size in (5.7), we again expect a nearly linear relationship between the maximum error and the step size in a loglog graph, but with a much steeper slope than before.

Comparing Euler, RK2, and RK4

It is very interesting to compare the accuracy of these three methods, both for individual step sizes, and for a range of step sizes.

Example 6. *Consider again the initial value problem*

$$y' = y + t, \quad y(0) = 1.$$

Sketch a graph of the maximum error in each of the three methods (Euler, RK2, RK4) versus the step size.

This can be done by writing a script M-file that is only a minor modification of batch3. Create a script M-file with the contents

```
h = 1; eul_vect = []; rk2_vect = []; rk4_vect = []; h_vect=[];
t = 0:.05:3;
y = -t - 1 + 2*exp(t);
for k=1:8
    [teuler,yeuler] = eul(@yplust,[0,3],1,h);
    [trk2,yrk2] = rk2(@yplust,[0,3],1,h);
    [trk4,yrk4] = rk4(@yplust,[0,3],1,h);
    plot(t,y,teuler,yeuler,'.',trk2,yrk2,'+',trk4,yrk4,'x')
    legend('Exact','Euler','RK2','RK4',2)
    shg
    zeuler = -teuler - 1 + 2*exp(teuler);
    eulerror = max(abs(zeuler - yeuler));
    zrk2 = -trk2 - 1 + 2*exp(trk2);
    rk2error = max(abs(zrk2 - yrk2));
    zrk4 = -trk4 - 1 + 2*exp(trk4);
    rk4error = max(abs(zrk4 - yrk4));
    h_vect = [h_vect,h]
    eul_vect = [eul_vect,eulerror]
    rk2_vect = [rk2_vect,rk2error]
    rk4_vect = [rk4_vect,rk4error]
    pause
    h = h/2;
end
loglog(h_vect,eul_vect,h_vect,rk2_vect,h_vect,rk4_vect)
legend('eul','rk2','rk4',4)
grid, xlabel('Step size'), ylabel('Maximum error')
title('Maximum error vs. step size'), axis tight
```

70

Save the file as `batch4.m`.

Most of the features of `batch4` are already present in `batch3`. The major difference is that we have added the plot commands for the loglog graph at the end, so that it will automatically be executed when the error data has been computed. Executing all eight steps in `batch4` will result in a plot of the error curves for each of the three methods, as shown in Figure 5.6.[4]

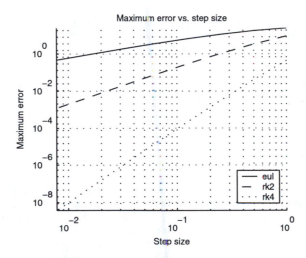

Figure 5.6. Error comparison for Euler, RK2, and RK4.

Note that Euler's method has the largest error for a given step size, with Runge-Kutta 2 being slightly better, while Runge-Kutta 4 is significantly better than the other two. Also notice that the curves are nearly straight lines in the loglog plot in Figure 5.6, indicating a power function relationship between the error and and the stepsize, as predicted by the inequalities (5.5), (5.6), and (5.7). The slopes of the curves are also roughly comparable to the order of the corresponding method. (Exercises 13–20 demonstrate the relation between a power function and its loglog plot. Exercise 21 experimentally determines the orders of the methods `eul`, `rk2`, and `rk4`)

Exercises

In Exercises 1–6 perform the following tasks for the given initial value problem on the specified interval.

a) Find the exact solution.

b) Use MATLAB to plot on a single figure window the graph of the exact solution, together with the plots of the solutions using each of the three methods (`eul`, `rk2`, and `rk4`) with the given step size h. Use a distinctive marker (type `help plot` for help on available markers) for each method. Label the graph appropriately and add a legend to the plot.

---

[4] As you execute `batch4.m` repeatedly, you will note that the vector `rk4_vect` appears to have zero entries except for the first two. Is the error from `rk4` actually zero at this point? The answer is no. To see what the entries really are we need to print the vector with more accuracy. To do this, enter `format long` and then `rk4_vect` to see what you get.

1. $x' = x\,\sin(3t)$, with $x(0) = 1$, on $[0, 4]$; $h = 0.2$.
2. $y' = (1 + y^2)\cos(t)$, with $y(0) = 0$, on $[0, 6]$; $h = 0.5$.
3. $z' = z^2\,\cos(2t)$, with $z(0) = 1$, on $[0, 6]$; $h = 0.25$.
4. $x' = x(4 - x)/5$, with $x(0) = 1$, on $[0, 6]$; $h = 0.5$.
5. $y' = 2ty$, with $y(0) = 1$, on $[0, 2]$; $h = 0.2$.
6. $z' = (1 - z)t^2$, with $z(0) = 3$, on $[0, 2]$; $h = 0.25$.

Using the code of Example 6 as a template, adjust the code to run with each of the initial value problems in Exercises 7–12. Note that you will first have to find an analytical solution for the given IVP. Observe how the approximations improve at each iteration and obtain a printout of the final loglog graph.

7. $y' = -y + 3t^2 e^{-t}$, with $y(0) = -1$, on $[0, 6]$.
8. $y' = 1 + 2y/t$, with $y(1) = 1$, on $[1, 4]$.
9. $y' = 3 - 2y/t$, with $y(1) = -1$, on $[1, 4]$.
10. $y' = y/t - y^2$, with $y(1) = 1/2$, on $[1, 4]$. *Hint: Look up Bernoulli's equation in your text.*
11. $y' = y + 2\cos t$, with $y(0) = -1$, on $[0, 2\pi]$.
12. $y' = 2te^t + y$, with $y(0) = -1$, on $[0, 2]$.

The function defined by the equation $y = a \cdot x^b$ is called a *power function*. You can plot the power function $y = 10x^2$ on the interval $[0.1, 2]$ with the commands x = linspace(0.1,2); y = 10*x.^2; plot(x,y). Take the logarithm base ten of both sides of $y = 10x^2$ to get $\log_{10} y = 1 + 2\log_{10} x$. Thus, if you plot $\log_{10} y$ versus $\log_{10} x$, the graph should be a line with slope 2 and intercept 1. You can create a loglog plot with the commands figure, plot(log10(x),log10(y)), grid on, then compare the slope and intercept of the result with the predicted result. Follow this procedure for each of the power functions in Exercises 13 –16. In each case, plot over the interval $[0.1, 2]$.

13. $y = x^3$
14. $y = 100x^4$
15. $y = 10x^{-2}$
16. $y = 1000x^{-5}$

MATLAB represents polynomials by placing the coefficients of the polynomial in a vector, always assuming that the polynomial is arranged in descending powers. For example, the vector p = [2,-5,3,-4] would be used to represent the polynomial $p(x) = 2x^3 - 5x^2 + 3x - 4$. The MATLAB function POLYFIT(X,Y,N) will fit a polynomial of degree N to the data stored in X and Y. Take the logarithm base ten of both sides of the power function $y = 10x^3$ to get $\log_{10} y = 1 + 3\log_{10} x$. As we saw in Exercises 13–16, the graph of $\log_{10} y$ versus $\log_{10} x$ is a line with slope 3 and intercept 1, plotted with the commands x = linspace(0.1,2); y = 10*x.^3; plot(log10(x), log10(y)), grid on. One can read the slope and intercept from the plot, or you can allow MATLAB to calculate them with p = polyfit(log10(x),log10(y),1), which responds with the polynomial p = [3 1]. The coefficients of this polynomial represent the slope and intercept, respectively, of $\log_{10} y = 3\log_{10} x + 1$. Exponentiating base ten, we recapture the power function $y = 10x^3$. In Exercises 17–20, follow this procedure to perform a loglog plot (base ten) of the given power function, then use the polyfit command to find the slope and intercept of the resulting line. Exponentiate the resulting equation to obtain the original power function.

17. $y = 10x^2$
18. $y = 100x^3$
19. $y = 10x^{-3}$
20. $y = 1000x^{-4}$

21. Execute batch4.m of Example 6 to regenerate the data sets and the plot shown in Figure 5.6. Now plot log10(rk4_vect) versus log10(h_vect) with the command plot(log10(h_vect),log10(rk4_vect), 'ro'). The points produced by this plot command should appear to maintain a linear relationship. Use the following code to determine and plot the "line of best fit" through the these data points (type help polyfit

and `help polyval` to get a full explanation of these commands).

```
hold on %hold the prior plot
p = polyfit(log10(h_vect),log10(rk4_vect),1)
plot(log10(h_vect),polyval(p,log10(h_vect)),'g')
```

Write the equation of the line of best fit in the form $\log_{10}(\text{rk4\_vect}) = C \log_{10}(\text{h\_vect}) + B$, where C and B are the slope and intercept captured from the `polyfit` command. Solve this last equation for `rk4_vect`. How closely does this last equation compare with the inequality (5.7)?

a) Perform a similar analysis on `eul_vect` and `h_vect` and compare with the inequality (5.5).

b) Perform a similar analysis on `rk2_vect` and `h_vect` and compare with the inequality (5.6).

22. Simple substitution will reveal that $x(t) = \pi/2$ is a solution of $x' = e^t \cos x$. Moreover, both $f(t, x) = e^t \cos x$ and $\partial f/\partial x = -e^t \sin x$ are continuous everywhere. Therefore, solutions are unique and no two solutions can ever share a common point (solutions cannot intersect). Use Euler's method with a step size $h = 0.1$ to plot the solution of $x' = e^t \cos x$ with initial condition $x(0) = 1$ on the interval $[0, 2\pi]$ and hold the graph (`hold on`). Overlay the graph of $x = \pi/2$ with the command `line([0,2*pi],[pi/2,pi/2],'color','r')` and note that the solutions intersect. Does reducing the step size help? If so, does this reduced step size hold up if you increase the interval to $[0, 10]$? How do `rk2` and `rk4` perform on this same problem?

23. The accuracy of any numerical method in solving a differential equation $y' = f(t, y)$ depends on how strongly the equation depends on the variable y. More precisely, the constants that appear in the error bounds depend on the derivatives of f with respect to y. To see this experimentally, consider the two initial value problems

$$y' = y, \qquad y(0) = 1,$$
$$y' = e^t, \qquad y(0) = 1.$$

You will notice that the two problems have the same solution. For each of the three methods described in this chapter compute approximate solutions to these two initial value problems over the interval $[0, 1]$ using a step size of $h = 0.01$. For each method compare the accuracy of the solution to the two problems.

24. Remember that $y(t) = e^t$ is the solution to the initial value problem $y' = y$, $y(0) = 1$. Then $e = e^1$, and in MATLAB this is `e = exp(1)`. Suppose we try to calculate e approximately by solving the initial value problem, using the methods of this chapter. Use step sizes of the form $1/n$, where n is an integer. For each of Euler's method, the second order Runge-Kutta method, and the fourth order Runge-Kutta method, how large does n have to be to get an approximation e_{app} which satisfies $|e_{app} - e| \le 10^{-3}$?

25. In the previous problem, show that the approximation to e using Euler's method with step size $1/n$ is $(1+1/n)^n$. As a challenge problem, compute the formulas for the approximations using the two Runge-Kutta methods.

Appendix: M-files for Numerical Methods

In this appendix we provide a listing of the programs `eul.m`, `rk2.m`, and `rk4.m` used in the chapter. The first listing is the file `eul.m`.

```
function [tout, yout] = eul(FunFcn, tspan, y0, ssize)
%
% EUL    Integrates a system of ordinary differential equations using
%    Euler's method.  See also ODE45 and ODEDEMO.
%    [t,y] = eul('yprime', tspan, y0) integrates the system
%    of ordinary differential equations described by the M-file
%    yprime.m over the interval tspan = [t0,tfinal] and using initial
%    conditions y0.
%    [t, y] = eul(F, tspan, y0, ssize) uses step size ssize
```

```
% INPUT:
% F       - String containing name of user-supplied problem description.
%            Call: yprime = fun(t,y) where F = 'fun'.
%            t       - Time (scalar).
%            y       - Solution vector.
%            yprime - Returned derivative vector; yprime(i) = dy(i)/dt.
% tspan = [t0, tfinal], where t0 is the initial value of t, and tfinal is
%            the final value of t.
% y0      - Initial value vector.
% ssize - The step size to be used. (Default: ssize = (tfinal - t0)/100).
%
% OUTPUT:
% t  - Returned integration time points (column-vector).
% y  - Returned solution, one solution row-vector per tout-value.

% Initialization
t0 = tspan(1); tfinal = tspan(2);
pm = sign(tfinal - t0);  % Which way are we computing?
if (nargin < 4), ssize = abs(tfinal - t0)/100; end
if ssize < 0, ssize = -ssize; end
h = pm*ssize;
t = t0; y = y0(:);
tout = t; yout = y.';

% We need to compute the number of steps.
dt = abs(tfinal - t0);
N = floor(dt/ssize) + 1;
if (N-1)*ssize < dt, N = N + 1; end

% Initialize the output.
tout = zeros(N,1);
tout(1) = t;
yout = zeros(N,size(y,1));
yout(1,:) = y.';
k = 1;

% The main loop
while k < N
  if pm*(t + h - tfinal) > 0
    h = tfinal - t;
    tout(k+1) = tfinal;
  else
    tout(k+1) = t0 +k*h;
  end
  k = k+1;
  % Compute the slope
  s1 = feval(FunFcn, t, y); s1 = s1(:); % s1 = f(t(k),y(k))
  y = y + h*s1;      % y(k+1) = y(k) + h*f(t(k),y(k))
  t = tout(k); yout(k,:) = y.';
end;
```

Since we are not teaching programming, we will not explain everything in this file, but a few things should be explained. The % is MATLAB's symbol for comments. MATLAB ignores everything that appears on a line after a %. The large section of comments that appears at the beginning of the file can be read using the MATLAB help command. Enter help eul and see what happens. Notice that these comments at the beginning take up more than half of the file. That's an indication of how easy it is to program these algorithms. In addition, a couple of lines of the code are needed only to allow the program to handle systems of equations.

Next, we present the M-file for the second order Runge-Kutta method, without the help file, which is pretty much identical to the help file in eul.m. You can type help rk2 to view the help file in MATLAB's command window.

```
function [tout, yout] = rk2(FunFcn, tspan, y0, ssize)

% Initialization
t0 = tspan(1);
tfinal = tspan(2);
pm = sign(tfinal - t0);   % Which way are we computing?
if nargin < 4, ssize = (tfinal - t0)/100; end
if ssize < 0, ssize = -ssize; end
h = pm*ssize;
t = t0;
y = y0(:);

% We need to compute the number of steps.
dt = abs(tfinal - t0);
N = floor(dt/ssize) + 1;
if (N-1)*ssize < dt
  N = N + 1;
end

% Initialize the output.
tout = zeros(N,1);
tout(1) = t;
yout = zeros(N,size(y,1));
yout(1,:) = y.';
k = 1;

% The main loop
while k < N
  if pm*(t + h - tfinal) > 0
    h = tfinal - t;
    tout(k+1) = tfinal;
  else
    tout(k+1) = t0 +k*h;
  end
  k = k + 1;
  s1 = feval(FunFcn, t, y); s1 = s1(:);
  s2 = feval(FunFcn, t + h, y + h*s1); s2 = s2(:);
  y = y + h*(s1 + s2)/2; t = tout(k);
  yout(k,:) = y.';
end;
```

Finally, we have the M-file for the fourth order Runge-Kutta method, without the help file. Type `help rk4` to view the help file in MATLAB's command window.

```
function [tout, yout] = rk4(FunFcn, tspan, y0, ssize)

% Initialization

t0 = tspan(1);
tfinal = tspan(2);
pm = sign(tfinal - t0);   % Which way are we computing?
if nargin < 4, ssize = (tfinal - t0)/100; end
if ssize < 0, ssize = -ssize; end
h = pm*ssize;
t = t0;
y = y0(:);


% We need to compute the number of steps.

dt = abs(tfinal - t0);
N = floor(dt/ssize) + 1;
if (N-1)*ssize < dt
  N = N + 1;
end

% Initialize the output.

tout = zeros(N,1);
tout(1) = t;
yout = zeros(N,size(y,1));
yout(1,:) = y.';
k = 1;

% The main loop
while (k < N)
  if pm*(t + h - tfinal) > 0
    h = tfinal - t;
    tout(k+1) = tfinal;
  else
    tout(k+1) = t0 +k*h;
  end
  k = k + 1;
  % Compute the slopes
  s1 = feval(FunFcn, t, y); s1 = s1(:);
  s2 = feval(FunFcn, t + h/2, y + h*s1/2); s2 = s2(:);
  s3 = feval(FunFcn, t + h/2, y + h*s2/2); s3 = s3(:);
  s4 = feval(FunFcn, t + h, y + h*s3); s4 = s4(:);
  y = y + h*(s1 + 2*s2 + 2*s3 +s4)/6;
  t = tout(k);
  yout(k,:) = y.';
end;
```

6.   Advanced Use of DFIELD6

In this chapter we will explore advanced uses of `dfield6`. In particular, we will experiment with differential equations that use step functions and square waves as forcing functions. We will also discover how to write function M-files for use in `dfield6`. In later sections we will learn how to use different solvers within `dfield6`.

Step Functions

The basic unit step function, called the *Heaviside* function, is defined as follows:

$$H(t) = \begin{cases} 0, & \text{if } t < 0; \\ 1, & \text{if } t \geq 0. \end{cases}$$

The easiest way to implement the Heaviside function in MATLAB is to use its *logical functions*. MATLAB assigns true statements a value of 1 and false statements a value of 0.

Confused? Try the following commands at the MATLAB prompt.

```
>> -2 < 0
ans =
     1
>> -2 >= 0
ans =
     0
```

In the first case, the inequality $-2 < 0$ is a true statement, so MATLAB outputs a 1. In the second case, $-2 \geq 0$ is a false statement, so MATLAB outputs a 0. A 1 means true, a 0 means false. The logical functions in MATLAB are array smart. For example,

```
>> t = -2:2, s = (t >= 0)
t =
    -2    -1     0     1     2
s =
     0     0     1     1     1
```

Notice that MATLAB assigns a value of 0 to every negative entry in the vector t, and a value of 1 to every nonnegative entry.

Example 1. *Use MATLAB to plot the Heaviside function.*

This is now easy. Simply execute

```
t = linspace(-2,2); plot(t,(t>=0)), shg
```

The result looks a little funny, but that is because the limits on the graph are not the best. Execute `axis([-2 2 -0.1 1.1])` and we will get something like Figure 6.1.

The Heaviside function is a "discontinuous" function, so the almost vertical line connecting the two constant portions of the graph in Figure 6.1 should not be present. It is present only because the MATLAB linestyle we selected connects consecutive points of the plot with line segments. If we replaced the `linspace` command with `linspace(-2,2,1000)`, the almost vertical portion would look truly vertical. It still should not be there, but we will put up with it.

The logical functions are particularly handy when a problem calls for the use of a *piecewise function*, a function that is defined in two or more pieces.

Example 2. *Sketch the graph of the piecewise defined function*

$$f(t) = \begin{cases} t, & \text{if } t < 1; \\ 1, & \text{if } t \geq 1; \end{cases} \tag{6.1}$$

on the interval $[0, 2]$.

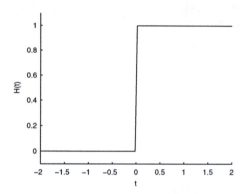

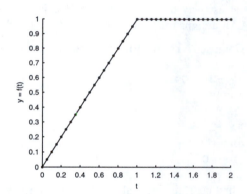

Figure 6.1. The Heaviside function.   **Figure 6.2.** The plot of f in (6.1).

Consider the following pseudocode, where we incorporate the use of MATLAB's logical functions,

$$g(t) = t \cdot (t < 1) + 1 \cdot (t \geq 1).$$

If $t < 1$, then $t < 1$ is a true statement and evaluates to 1, but $t \geq 1$ is a false statement and evaluates to 0. Consequently, if $t < 1$,

$$g(t) = t \cdot (t < 1) + 1 \cdot (t \geq 1) = t \cdot (1) + 1 \cdot (0) = t.$$

However, if $t \geq 1$, then $t < 1$ is false and evaluates to 0, but $t \geq 1$ is now a true statement and evaluates to 1. Consequently, if $t \geq 1$,

$$g(t) = t \cdot (t < 1) + 1 \cdot (t \geq 1) = t \cdot (0) + 1 \cdot (1) = 1.$$

Note that these results agree with the definition of $f(t)$ in equation (6.1). Consequently, $f(t) = g(t)$ for all values of t and we can use $f(t) = t \cdot (t < 1) + 1 \cdot (t \geq 1)$ to sketch the graph of the piecewise function $f(t)$.

Using the logical functions we execute

```
>> t = 0:.05:2;
>> y = t.*(t<1) + 1*(t>=1);
>> plot(t,y,'.-')
```

This second line warrants some explanation. Because `t` and `t<1` are vectors, array multiplication `.*` is required in the product `t.*(t < 1)`. However, array multiplication is not required in the scalar-vector product `1*(t >= 1)` (although we could also use `1.*(t >= 1)` — try it!). The graph is shown in Figure 6.2.

Step Functions in DFIELD6

Now that we know how to implement step functions in MATLAB, let's try to use them in applications requiring the use of `dfield6`.

Example 3. *The voltage V_c across the capacitor in an electric circuit containing a resistor and capacitor (an RC-circuit) satisfies the ODE*

$$RC\frac{dV_c}{dt} + V_c = V(t),\qquad(6.2)$$

where R is the resistance, C is the capacitance, and $V(t)$ is an external voltage source. Suppose that we have a voltage source of the form

$$V(t) = \begin{cases} 2, & \text{if } t < 3; \\ 0, & \text{if } t \geq 3. \end{cases}\qquad(6.3)$$

Assume that the initial voltage across the capacitor is $V_c(0) = 0$ volts, that the resistance is $R = 0.2$ ohms, the capacitance is $C = 1$ farad, and time is measured in seconds. Use `dfield6` to sketch the solution of this initial value problem over the time interval $[0, 6]$.

The first step is to put the ODE (6.2) into normal form,

$$\frac{dV_c}{dt} = \frac{V(t) - V_c}{RC}.\qquad(6.4)$$

This is the equation we want to enter in the DFIELD6 Setup window.

The driving voltage, $V(t)$, in equation (6.3) is "on" at a level of 2 volts for the first three seconds of the time interval $[0, 6]$, but "off" for the remaining three seconds of this time interval. We will enter V as an expression. Pseudocode for $V(t)$ is $2 \cdot (t < 3) + 0 \cdot (t \geq 3)$, or, more simply, $V(t) = 2 \cdot (t < 3)$. Entering all of this information into the DFIELD6 Setup window results in Figure 6.3. Notice that we are using TEX notation for the subscripted V_c.

We use the DFIELD6 Keyboard input dialog box to compute the trajectory with initial condition $V_c(0) = 0$, as shown in Figure 6.4. It is interesting to interpret the meaning of the trajectory shown in Figure 6.4. As the power source is turned "on" (with an amplitude of 2 volts), the voltage across the capacitor increases to a level of 2 volts. At $t = 3$ seconds, the power is turned "off" (0 volts) and the voltage across the capacitor decays exponentially to zero.

79

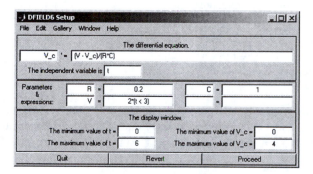

Figure 6.3. Setting up the circuit equation.

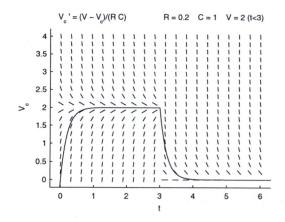

Figure 6.4. Solution trajectory with initial condition $V_c(0) = 0$.

Using Function M-files in DFIELD6

Let's look at another example that exhibits new features of dfield6.

Example 4. *Suppose that originally there are 10 tons of fish in a lake. The fish population is governed by the logistic equation with a natural reproductive rate of $r = 1$ and a carrying capacity of $K = 10$. Starting at $t = 0$, the fish are harvested at the rate of 3 tons per month for the first three months and 0.5 tons per month after that. Use* dfield6 *to sketch the population over 18 months.*

We will use tons as the units for P and months for t. Then the population $P(t)$ satisfies the equation

$$\frac{dP}{dt} = rP\left(1 - \frac{P}{K}\right) - H, \tag{6.5}$$

where the harvesting function H is defined by

$$H(t) = \begin{cases} 3, & \text{if } t \le 3, \\ 0.5, & \text{if } t > 3. \end{cases} \tag{6.6}$$

80

Note that the harvesting strategy in equation (6.6) exhibits heavy harvesting for the first three months (3 tons per month), while the harvesting is lighter in the remaining months (0.5 tons per month).

We could handle this problem using the same method we used in Example 3, by entering the right-hand side of equation (6.5) in the DFIELD6 Setup window, and then using the pseudocode $H(t) = 3 \cdot (t \le 3) + 0.5 \cdot (t > 3)$ to define the expression H. However, we want to use this example to illustrate the use of function M-files in dfield6.

Start by opening your editor and creating a function M-file with the following code.

```
function H = harvest(t)
if t <= 3
    H = 3;
else
    H = 0.5;
end
```

At first glance this looks like the perfect M-file for our harvesting function $H(t)$. However, function M-files used in dfield6 must be array smart, and the function harvest is not. We can see this by testing it, say on the vector 0:6. To make harvest array smart, we use the fact that the function is piecewise defined and use the logical functions as we have been doing. A correct version is

```
function H = harvest(t)
H = 3*(t <= 3) + 0.5*(t > 3);
```

Now we get

```
>> harvest(0:6)
ans =
    3.0000   3.0000   3.0000   3.0000   0.5000   0.5000   0.5000
```

so the function is array smart.

Enter the differential equation in (6.5) into the DFIELD6 Setup window. Set the parameters K and r, and use H = harvest(t) as an expression. After setting the display window dimensions, the result is Figure 6.5. Use the DFIELD6 Keyboard input box to compute the trajectory with initial condition $P(0) = 10$, and we get a plot similar to that shown in Figure 6.6. Notice that the population starts at the initial condition $P(0) = 10$, which is the carrying capacity. It then declines rapidly because of the heavy fishing. At 3 months, when the level of fishing is decreased, it begins to increase, but it levels off at a point somewhat lower than the original carrying capacity.

The previous example illustrates the use of a simple function M-file in dfield6. This feature becomes more useful when more complicated functions are used.

The Square Wave Function

Let's digress and discuss a function that is important in science and engineering, the *square wave* function. The square wave is a periodic function of period T. During a certain percentage of its period, called the *duty cycle d*, the square wave function is "on," meaning that it is equal to 1. During the remaining portion of its period it is "off," or equal to 0.

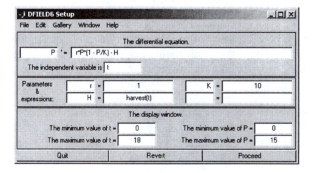

Figure 6.5. Setting up Example 4.

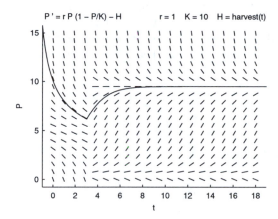

Figure 6.6. Solution with initial condition $P(0) = 10$.

Let's write a function M-file to implement the square wave function. The one new idea we need is implemented in the MATLAB function mod. You can use `help mod` to get more information about it, but the operation is easily understood as "remainder after division." Given any two numbers t and T, with $T > 0$, we can write $t = n \cdot T + r$, where the *quotient* n is an integer and the *remainder* r satisfies $0 \le r < T$. Then `mod(t,T)` = r.

For the square wave, which we will denote by sqw, we are given a period T and the duty cycle d. Notice that d is a percentage. Hence, we have

$$\text{sqw}(t, T, d) = \begin{cases} 1, & \text{if } r < dT/100; \\ 0, & \text{otherwise.} \end{cases} \quad , \quad \text{where} \quad r = \text{mod}(t, T).$$

Thus the square wave is defined as a piecewise function of the remainder $r = \text{mod}(t, T)$, so we can

82

implement it using MATLAB's logic functions. Here is a correct function M-file.

```
function y = sqw(t,T,d)
r = mod(t,T);
y = (r < d*T/100);
```

Since we will use the square wave function repeatedly, and since it is not a standard part of MATLAB, we suggest that you create this M-file and store it in a directory on your MATLAB path, so that it will be available to you when you want it.

The following code will produce the graph of a square wave of period $T = 4$ and duty cycle $d = 25$ in a new figure window as shown in Figure 6.7.[1]

```
>> figure
>> t = linspace(0,12,1000);
>> y = sqw(t,4,25);
>> plot(t,y)
>> set(gca,'xtick',0:12)
```

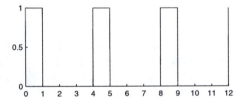

Figure 6.7. Square wave: Period = 4, Duty = 25%

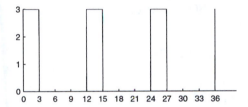

Figure 6.8. The harvesting strategy is periodic

The square wave in Figure 6.7 has period 4 and is "on" (equal to 1) for the first 25% of its period, then "off" (equal to 0) for the remainder. Setting "tick marks" at $0, 1, 2, \ldots, 12$ makes this behavior clearer. This is accomplished with the handle graphics command set, which we explained at the end of Chapter 2.[2] The plot clearly illustrates how the square wave function got its name.

Since the square wave is discontinuous, the vertical lines in Figure 6.7 are only present because the MATLAB linestyle we selected connects consecutive points of the plot with line segments. We used 1000 points in the linspace command to make the lines look truly vertical.

---

[1] The plot command always uses the current figure window. Clicking on any figure window with the mouse makes it the current figure window. MATLAB's figure command creates a new figure window and makes it the current figure window.

[2] If you are running version 6.5 of MATLAB, you can also adjust the tick marks using the Axes Properties Editor found in the **Edit** menu on your figure window. Select the "X" tab, then uncheck "Auto" on the "Ticks" checkbox. This will allow you to enter 0:12 in the edit box. Select "Apply," then "OK" to edit the axes properties dialog window. Use your mouse to click the Arrow icon on the toolbar to complete the process.

Example 5. *Consider again the fish population in Example 4. This time let's propose a different harvesting function $H(t)$. Let's suppose that harvesting is performed at a constant rate of 3 tons per month for the first three months of the year, but no harvesting is done for the remainder of the year, and that this pattern repeats every year. Use* dfield6 *to plot the population over a three year period.*

Once again, the differential equation is

$$\frac{dP}{dt} = rP\left(1 - \frac{P}{K}\right) - H, \tag{6.7}$$

but we have to figure out how to implement the harvesting function H.

Our harvesting strategy calls for three tons per month for the first three months of the year, then zero harvesting the remainder of the year. Note that three months is precisely 25% of a year — so the duty cycle is 25%. We want to examine the fish population over three years (36 months). The following code is used to produce the harvesting function pictured in Figure 6.8.

```
>> t = linspace(0,36,1000);
>> y = 3*sqw(t,12,25);
>> plot(t,y)
>> set(gca,'xtick',0:3:36)
```

Multiplying sqw by three is all that is needed to change the amplitude of the square wave. The square wave depicted in Figure 6.8 is exactly the harvesting strategy required by Example 4. Fish are harvested at a rate of three tons per month for the first three months of each year, and this pattern repeats itself every 12 months. Thus $H(t) = 3\,\mathrm{sqw}(t, 12, 25)$. With $0 \le t \le 36$ and $0 \le P \le 15$, enter all of the data into the DFIELD6 Setup window as shown in Figure 6.9.

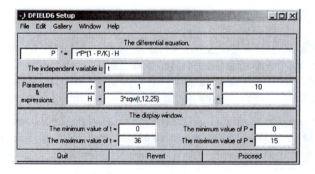

Figure 6.9. Setting up the differential equation.

Use the Keyboard Input window to start a solution trajectory with initial condition $P(0) = 10$. The result is the top graph in Figure 6.10. It is revealing to superimpose the square wave of Figure 6.8 onto this figure, and to change the tick marks as well. However, we want the square wave to be plotted over the entire range displayed in Figure 6.10. To do this we change the first command to linspace(-3,39,1000) in the previous list, and execute all four commands, making sure that the DFIELD6 Display window is the active figure.

So, what is the fate of the fish population? When you examine Figure 6.10, the fish population starts at 10 tons, then decays somewhat rapidly during the harvest season (the first three months), then makes a slow recovery to 10 tons over the next nine months. This pattern then repeats itself over the next two years.

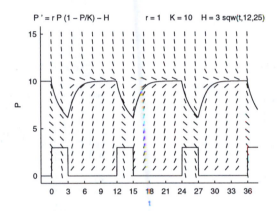

Figure 6.10. The fate of the fish population.

Solvers and Solver Settings in DFIELD6

Dfield6 makes available a number of important solvers. To view the various solvers available in dfield6, select **Options→Solver**. The default solver is the Dormand-Prince algorithm.[3] It is a variable step algorithm that adjusts the step size to achieve a prescribed error tolerance. The solvers Euler, Runge-Kutta 2, and Runge-Kutta 4 are identical to the routines eul.m, rk2.m, and rk4.m discussed in Chapter 5. Each of these is a fixed or constant-step algorithm and the user must fix the step size in advance.

We can best demonstrate the use of these solvers with an example.

Example 6. *Consider an RC-circuit like that in Example 3 with resistance $R = 2.3$ ohms and capacitance $C = 1.2$ farads. Assume the driving voltage is a square wave with amplitude 3 volts, period 1 second, and duty cycle 10%. Consequently, equation (6.4) becomes*

$$V_c' = \frac{V - V_c}{RC} \quad \text{where} \quad V = 3\,\text{sqw}(t, 1, 10).$$

Sketch the solution with initial condition $V_c(0) = 0$.

---

[3] This is the same algorithm used in ode45, one of MATLAB's suite of solvers. We will discuss ode45 in Chapter 8.

Enter the data as we did in Example 3, but set the value for the expression V to `3*sqw(t,1,10)`. Set the display window to $0 \le t \le 24$ and $0 \le V_c \le 0.5$. Click **Proceed** to transfer information to the DFIELD6 Display window.

We want to be sure that we are using the default solver and settings. If you have not changed these since starting `dfield6`, you need do nothing. Otherwise, select **Options→Solver→Dormand-Prince**. Reset the Number of plot steps per computation step to 4 and the relative error tolerance to 1×10^{-6} (`1e-006`) in the DFIELD6 Solver settings window. Double click the **Change settings** button.[4] Use the Keyboard input window to start the solution trajectory with initial condition $V_c(0) = 0$. The result is the graph in Figure 6.11.

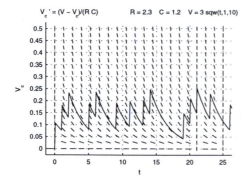

Figure 6.11. The Dormand-Prince solution.

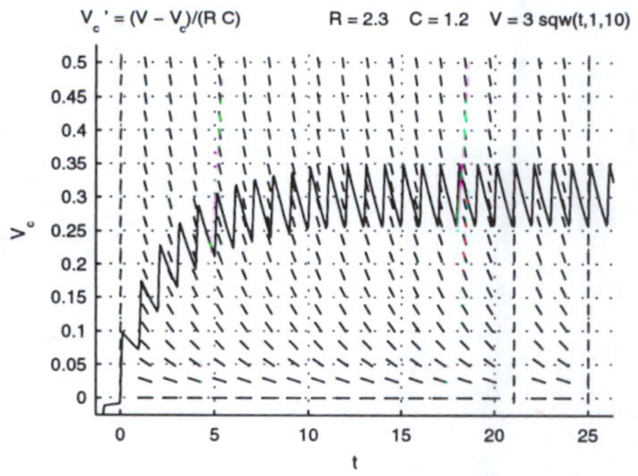

Figure 6.12. The Runge-Kutta 4 solution.

This solution cannot be correct! Since the driving voltage is a square wave, periodic with period 1, we expect the capacitor to charge and discharge in each period of length one second. That is not what we see. The problem lies with the fact that an adaptive (variable-step) solver, like Dormand-Prince, might take too large a step-size, completely bypassing the point where the input voltage modeled by the square wave suddenly changes from zero to three volts.

There is a basic reason why the Dormand-Prince solver is a bad choice for this example. Every high order solver is highly accurate, but only when the right-hand side of the equation is a nice, smooth function. Our square wave is discontinuous, so we should expect problems. We can avoid the difficulties if we set things up so that there are solution points at the points of discontinuity of the right-hand side. This is hard to arrange with a variable step solver, but it is not too hard with a fixed step solver like Runge-Kutta. In this case the square wave function with period 1 and duty 10% has discontinuities at t = 0, 0.1, 1.0, 1.1, ... To be sure that all of these are step points we need a step size which is an integer fraction of 0.1, say 0.05.

To implement this, select **Options→Solver→Runge-Kutta 4**. In the DFIELD6 Solver settings window set the step size to 0.05. Use the Keyboard input window to start the solution trajectory with

---

[4] Due to a bug in the MATLAB software, the **Change settings** button may not enable. If you double click it, it will enable after the first click and then take action after the second.

initial condition $V_c(0) = 0$. If you've followed these directions carefully, your DFIELD6 Display window should resemble that shown in Figure 6.12.

We can try to address the problem using the Dormand-Prince solver. Select **Options→Solver→ Dormand-Prince**. Then select **Options→Solver settings** and change the **maximum step size**. The default is 2.4, which is by default 1/10 of the time interval set in the DFIELD6 Setup window. Just as we chose 0.05 as a suitable step size for Runge-Kutta, we set the maximum step size to 0.05 and click the **Change settings** button. If the maximum step size is smaller than every variable step size computed by the solver, the solver will act like a fixed step solver, and everything will be OK. In general, this is risky, but it works in the case at hand. Use the Keyboard input dialog box to restart a solution trajectory with initial condition $V_c(0) = 0$. Note that this solution trajectory appears to closely mimic the Runge-Kutta 4 solution trajectory.

Another comment is in order. When you begin adjusting the maximum step size of a variable-step solve like Dormand-Prince, you are essentially changing its character to that of a fixed step Runge-Kutta solver. It is wiser to choose a solver appropriate for the job at hand, like rk4 in this case.

Kinky plots. Should you get a plot that appears "kinky" or "jagged," and you have no discontinuities such as we saw in Example 6, you can adjust the number of plot steps per computation step to smooth your solution trajectory. The Dormand-Prince algorithm performs interpolation between computation steps. This is accomplished without greatly degrading speed or the efficiency of the algorithm. To change the number of interpolated points between computation steps, select **Options→Solver settings** to open the DFIELD6 Solver settings window, then increase the **Number of plotted points per computation step**. This should smooth most jagged-looking curves, unless of course, they are supposed to be jagged.

Error tolerance and step size. You might have need to adjust the accuracy of a solution. To do so, Select **Options→Solver settings**. This opens the DFIELD6 Settings window, which has different options for the different solvers. If you are using the Dormand Prince solver, you can set the relative error tolerance and the maximum step size to what you want If you are using one of the fixed step size solvers, you can set the step size. Double click the **Change settings** button to make the changes effective.

Exercises

1. Use your editor to create a Heaviside function.

```
function y = Heaviside(t)
y = (t >= 0);
```

   Use this Heaviside function to obtain a plot of the Heaviside function on the interval $[-3, 3]$. Experiment with different linestyles and obtain a printout of your favorite.

2. You can use the Heaviside function defined in Exercise 1 to create a number of interesting plots.
   a) Use the command sequence t = linspace(-3,3,1000); y = Heaviside(t-1); plot(t,y,'o') to create a "delay" of one second in turning "on" the switch represented by the Heaviside function.
   b) Use the command sequence t = linspace(-3,3,1000); y = Heaviside(-t); plot(t,y,'o') to reflect the usual graph of the Heaviside function across the vertical axis.
   c) Use the command sequence t = linspace(-3,3,1000); y = Heaviside(t + 1) - Heaviside(t - 1); plot(t,y,'o') to create a "pulse" of width two and height one, centered at the origin.

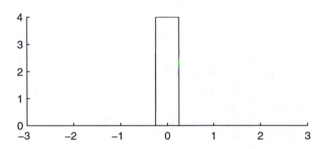

Figure 6.13. A pulse function

d) What sequence of commands will produce the pulse shown in Figure 6.13?

3. Obtain a graph of the plot produced by the command sequence

```
t = linspace(0,3);
y = t.*(t < 1) + 1*(t >= 1 & t < 2) + t.^2.*(t >= 2);
plot(t,y)
```

For each of the function in Exercises 4 – 7, write an array smart function M-file, and plot the function over the indicated interval. For Exercises 6 and 7 there is a hint in Exercise 3,

4. The *ramp function* $R(t) = \begin{cases} 0, & \text{if } t < 0, \\ t, & \text{if } t \ge 0, \end{cases}$ on the time interval $[-3, 3]$.

5. $f(t) = \begin{cases} -t, & \text{if } t < 0, \\ t^2, & \text{if } t \ge 0, \end{cases}$ on the time interval $[-2, 2]$.

6. $f(t) = \begin{cases} 1, & \text{if } t < 1, \\ 2, & \text{if } 1 \le t < 2, \\ 3, & \text{if } t \ge 2, \end{cases}$ on the interval $[0, 3]$.

7. $h(t) = \begin{cases} 0, & \text{if } t < 0, \\ t, & \text{if } 0 \le t < 2, \\ 2, & \text{if } 2 \le t < 4, \\ 6 - t, & \text{if } t \ge 4, \end{cases}$ on the time interval $[-1, 5]$.

8. An *RC*-circuit is modeled by the equation

$$RC\frac{dV_c}{dt} + V_c = V(t),$$

where V_c is the voltage across the capacitor. Suppose that the resistance $R = 2.3$ ohms, the capacitance $C = 1.2$ farads, and the driving voltage of the circuit is supplied by

$$V(t) = \begin{cases} 3, & \text{if } t < 5; \\ 0, & \text{if } t \ge 5. \end{cases}$$

Use dfield6 to sketch the solution of the circuit with initial condition $V_c(0) = 0$.

9. Use the circuit of Exercise 8, with the same resistance and capacitance, but this time the driving voltage is given by

$$V(t) = \begin{cases} 0, & \text{if } t < 5; \\ 3, & \text{if } t \ge 5. \end{cases}$$

Use `dfield6` to sketch the solution of the circuit with initial condition $V_c(0) = 2$. What appears to be the eventual voltage V_c across the capacitor?

10. Create the following function M-file and save it as `voltage.m`.

```
function v = voltage(t)
v = 3*(t >= 5);
```

Enter the differential equation of Exercises 8 and 9 in the DFIELD6 Setup window as `V_c' = (voltage(t) - V_c)/(R*C)`. Use this equation to duplicate the result in Exercise 9.

11. Assume time is measured in seconds. Use the function M-file `sqw` to obtain plots of a square wave on the time interval [0, 24]

    a) with period T=3 and duty cycle 25%.

    b) with period T=2 and duty cycle 50%.

    c) with period T=4 and duty cycle 75%.

    d) Create a plot of a square wave that is "on" for four seconds, "off" for eight seconds, "on" for four seconds, "off" for eight seconds, etc. Obtain a printout of your result on the time period [0, 96].

12. Examine the plot generated by

```
>> t = linspace(0,24,1000);
>> y = sqw(-t,8,50);
>> plot(t,y,'o-')
```

Explain how to use the `sqw` function to obtain a square wave of period four seconds that is "off" for the first second of each period, and "on" for the remaining three seconds in each period. Obtain a printout of your result on the time interval [0, 24].

13. Consider again the RC-circuit of Exercise 8, but this time the voltage source is an amplified square wave, with a period of thirty-two seconds and a duty cycle of 25%. Enter `V_c' = (3*sqw(t,32,25) - V_c)/(R*C)` in the DFIELD6 Setup window. Keep the parameters `R = 2.3` and `C = 1.2`, and adjust the display window so that $0 \leq t \leq 96$ and $0 \leq V_c \leq 4$.

    a) Start a solution trajectory with initial condition $V_c(0) = 0$.

    b) Make the DFIELD6 Display window inactive with **Options→Make the Display Window inactive**. Enter the following code in the MATLAB command window to superimpose the input voltage on the DFIELD6 Display window.

```
>> t = linspace(0,96,1000);
>> y = 3*sqw(t,32,25);
>> plot(t,y)
```

Think of the driving voltage as a signal input into the circuit and the solution trajectory as the resulting output signal. Note that the input signal has been somewhat *attenuated*; that is, the amplitude has decreased in the output signal.

    c) Lower the period of the square wave input by entering `V_c' = (3*sqw(t,16,25) - V_c)/(R*C)` in the DFIELD6 Setup window. Start a solution trajectory with initial condition $V_c(0) = 0$. Superimpose the square wave as you did in part b). Note that the output signal is even further attenuated. Repeat this exercise, using periods of 8 and 4 for the input voltage and note how the circuit attenuates input signals with smaller periods (higher frequencies). Why would engineers name this circuit a low pass filter?

14. The plot shown in Figure 6.14 is called a *sawtooth wave*. It has period six and amplitude three. Figure 6.14

was produced with the following MATLAB commands.

```
>> t = linspace(0,24,1000);
>> y = 3*mod(t,6)/6;
>> plot(t,y)
>> set(gca,'xtick',0:3:24)
```

Draw sawtooth waves

a) with period 3 and amplitude 1.

b) with period 4 and amplitude 3.

c) with period 8 and amplitude 4.

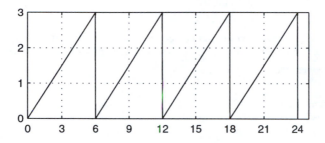

Figure 6.14. A sawtooth wave with period 6, amplitude 3.

15. Consider again the RC-circuit of Exercise 8, but this time the voltage source is an amplified sawtooth wave, with a period of thirty two seconds. Enter V_c' = (3*mod(t,32)/32 - V_c)/(R*C) in the DFIELD6 Setup window. Keep the parameters R = 2.3 and C = 1.2, and adjust the display window so that $0 \le t \le 96$ and $0 \le V_c \le 4$.

a) Start a solution trajectory with initial condition $V_c(0) = 0$.

b) Make the DFIELD6 Display window inactive. Enter the following code in the MATLAB command window to superimpose the input voltage on the DFIELD6 Display window.

```
>> t = linspace(0,96,1000);
>> y = 3*mod(t,32)/32;
>> plot(t,y)
```

c) Lower the period of the sawtooth wave input by entering V_c' = (3*mod(t,16)/16 - V_c) /(R*C) in the DFIELD6 Setup window. Start a solution trajectory with initial condition $V_c(0) = 0$. Superimpose the sawtooth wave as you did in part b). Repeat this exercise, using periods of 8 and 4 for the input voltage. Which signals are least attenuated by this circuit?

16. Consider again the fish population from Example 4, where $dP/dt = (1 - P/10)P - H(t)$ described the harvesting of a fish population and the time was measured in months. The harvesting strategy was defined by the piecewise function

$$H(t) = \begin{cases} 3, & \text{if } t \le 3, \\ 0.5, & \text{if } t > 3, \end{cases}$$

which modeled heavy harvesting in the first three months of each year, lighter harvesting during the remaining months of the year.

90

The fish population rebounded in Example 4, but this is not always the case. Use the **Keyboard Input** box to help find the critical initial fish population that separates life from extinction.

17. Consider a fish population with constant harvesting, modeled by the equation $dP/dt = (1 - P/10)P - H$. If there is no harvesting allowed, then the fish population behaves according to the logistic model (See Example 5 in Chapter 3).

   a) Enter the differential equation in the DFIELD6 Setup window. Add a parameter H and set it equal to zero. Use the **Keyboard input** box to sketch the equilibrium solutions. The equilibrium solutions divide the plane into three regions. Sketch one solution trajectory in each of the three regions.

   b) Slowly increase the parameter H and note how the equilibrium solutions move closer together, eventually being replaced by one, then no equilibrium solutions. This is called a *bifurcation* and has great impact on the eventual fate of the fish population. Use dfield6 to experimentally determine the value of H at which this bifurcation takes place. *Note:* If you find it difficult to find the exact bifurcation point with dfield6, you might consider the graph of $(1 - P/10)P - H$ versus P, as in Figure 3.16 in Chapter 3. For what value of H does this plot have exactly one intercept? Check your response in dfield6.

   c) Prepare a report on the effect of constant harvesting on this fish population. Your report should include printouts of the DFIELD6 Display window for values of H on each side of the bifurcation point and an explanation of what happens to the fish population for a variety of different initial conditions.

18. One can easily advance or delay a square wave. For example, the MATLAB commands

```
>> t = linspace(0,12,1000);
>> y = sqw(t-1,4,25);
>> plot(t,y)
>> set(gca,'xtick',0:12)
```

produce a square wave with period T=4 and duty cycle d=25%, but the wave is "delayed" one second, as shown in Figure 6.15.

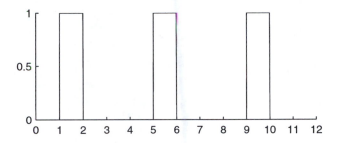

Figure 6.15. A square wave that is delayed one second.

Some might find the term "delay" misleading, as the square wave appears to be moved *forward* one second in time. However, the square pulse does start one second later than usual, hence the term "delay." Use MATLAB to create a plot of a square wave with

   a) amplitude three, period $T = 6$ seconds, duty cycle $d = 25\%$, that is delayed 2 seconds on the time interval $[0, 24]$.

   b) amplitude four, period $T = 12$ months, duty cycle $d = 25\%$, that is delayed 9 months on the time interval $[0, 48]$.

19. Consider again the harvest of the fish population of Example 5, modeled by $dP/dt = (1 - P/10)P - H(t)$, where the time is measured in months. In Example 5, the strategy called for a constant harvesting rate of 3

tons per month, but only during the months of January, February, and March, with no harvesting allowed in the remaining months of the year.

a) Suppose that the harvesting season is actually during March, April, and May. Use `dfield6` to show the fate of an initial population of ten tons of fish that are harvested at a constant rate of 3 tons per month during harvesting season, but not at all during the remaining months of the year. *Hint:* See Exercise 18.

b) Show the fate of the population in part a) if the harvesting season is in November, December, and January of each year.

c) In part b), the fate of the fish population depends on the initial population of fish. Use `dfield6` to estimate the critical initial condition that separates survival from extinction.

7.   Introduction to PPLANE6

A *planar system* is a system of differential equations of the form

$$x' = f(t, x, y),$$
$$y' = g(t, x, y). \tag{7.1}$$

The variable t in system (7.1) usually represents time and is called the *independent* variable. Frequently, the right-hand sides of the differential equations do not explicitly involve the variable t, and can be written in the form

$$x' = f(x, y),$$
$$y' = g(x, y). \tag{7.2}$$

Such a system is called *autonomous*.

The system

$$x' = y,$$
$$y' = -x, \tag{7.3}$$

is an example of a planar autonomous system. The reader may check, by direct substitution, that the pair of functions $x(t) = \cos t$ and $y(t) = -\sin t$ satisfy both equations and is therefore a solution of system (7.3).

There are a number of ways that we can represent the solution of (7.3) graphically. The following code was used to construct the plots of x and y versus t in Figure 7.1.

```
>> t = linspace(0,4*pi);
>> x = cos(t); y = -sin(t);
>> subplot(2,1,1)
>> plot(t,x)
>> title('x = cos t'), ylabel('x')
>> subplot(2,1,2)
>> plot(t,y)
>> title('y = -sin t')
>> xlabel('t'), ylabel('y')
```

You can also plot the solution in the so-called *phase plane*. The commands

```
>> plot(x,y)
>> axis equal
>> title('The plot of y versus x')
>> xlabel('x'), ylabel('y')
```

produce a plot of y versus x in the phase plane, as shown in Figure 7.2.

System (7.1) can be written as the vector equation

$$\begin{bmatrix} x \\ y \end{bmatrix}' = \begin{bmatrix} x' \\ y' \end{bmatrix} = \begin{bmatrix} f(t, [x, y]^T) \\ g(t, [x, y]^T) \end{bmatrix}. \tag{7.4}$$

93

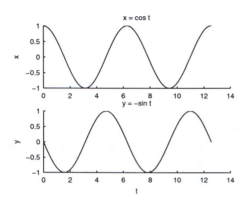

Figure 7.1. Plots of x and y versus t.

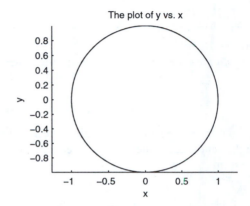

Figure 7.2. The solution in the phase plane.

If we let $\mathbf{x} = [x, y]^T$, then $\mathbf{x}' = [x', y']^T$ and equation (7.4) becomes

$$\mathbf{x}' = \begin{bmatrix} f(t, \mathbf{x}) \\ g(t, \mathbf{x}) \end{bmatrix}. \tag{7.5}$$

Finally, if we define $\mathbf{F}(t, \mathbf{x}) = [f(t, \mathbf{x}), g(t, \mathbf{x})]^T$, then equation (7.5) can be written as

$$\mathbf{x}' = \mathbf{F}(t, \mathbf{x}). \tag{7.6}$$

For a general planar system of the form $\mathbf{x}' = \mathbf{F}(t, \mathbf{x})$, a solution is a vector valued function $\mathbf{x}(t) = [x(t), y(t)]^T$. The individual components can be plotted as in Figure 7.1, or you can plot in the phase plane as shown in Figure 7.2. The fact that the function $\mathbf{x}(t)$ is a solution of the differential equation $\mathbf{x}' = \mathbf{F}(t, \mathbf{x})$ means that at every point $(t, \mathbf{x}(t))$, the curve $t \to \mathbf{x}(t)$ must have $\mathbf{F}(t, \mathbf{x}(t))$ as a tangent vector. For a fixed value of t we can imagine the vector $\mathbf{F}(t, \mathbf{x})$ attached to the point $\mathbf{x}$, representing the collection of all possible tangent vectors to solution curves for that specific value of t. Unfortunately this vector field changes as t changes, so this rather difficult visualization is not too useful. However, if the system is autonomous, then system (7.6) can be written

$$\mathbf{x}' = \mathbf{F}(\mathbf{x}) \tag{7.7}$$

and the vector field $\mathbf{F}(\mathbf{x})$ does not change with time t. Therefore, for an autonomous system the same vector field represents all possible tangent vectors to solution curves for all values of t. If a solution curve is plotted parametrically, at each point the vector field must be tangent to the solution curve.

The MATLAB function pplane6 makes this visualization easy.[1] This chapter provides an introduction to pplane6. We will delay discussing more advanced features of pplane6 until you have learned more about systems of differential equations in the ensuing chapters. Actually, the functionality of pplane6 is very similar to that of dfield6, so if you are familiar with dfield6, you will have no trouble with pplane6.

---

[1] To see if pplane6 is installed properly on your computer enter help pplane6. If it is not installed see the Appendix to Chapter 3 for instructions on how to obtain it.

Starting PPLANE6

To see `pplane` in action, enter `pplane6` at the MATLAB prompt. A new window will appear with the label PPLANE6 Setup. Figure 7.3 shows how this window looks on a PC running Windows. The appearance might differ slightly depending on your computer, but the functionality will be the same on all operating systems.

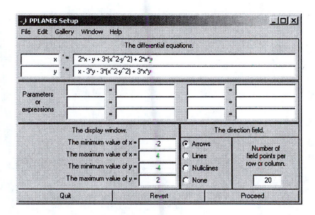

Figure 7.3. The setup window for `pplane6`.

You will notice that there is a rather complicated autonomous system already entered in the upper part of the PPLANE6 Setup window, in the form $x' = f(x, y)$ and $y' = g(x, y)$ of equation (7.2). There is a middle section for entering parameters and expressions, although none are entered at the moment. There is another section for describing a "display window," and yet another for defining what kind of field is to be used. There are are three buttons at the bottom of the window and several menus across the top. These are the same buttons and menus that are found in the DFIELD6 Setup window and they work in `pplane6` just as they do in `dfield6`. We will describe the use of the menus in detail later, but for now leave everything unchanged and click the button labeled **Proceed**. After a few seconds another window will open, this one labeled PPLANE6 Display. An example of this window is shown in Figure 7.4.

The most prominent feature of the PPLANE6 Display window is a rectangle labeled with the variable x on the bottom, and the variable y on the left. The dimensions of this rectangle are slightly larger than the rectangle specified in the PPLANE6 Setup window in order to accommodate the extra space needed by the vectors in the field. Inside this rectangle the PPLANE6 Display window shows the vector field for the system defined in the PPLANE6 Setup window. At each point (x, y) of a grid of points, there is drawn a small arrow. The direction of the vector is the same as that of $\mathbf{F}(x, y) = [f(x, y), g(x, y)]^T$, entered as differential equations in the PPLANE6 Setup window, and the length varies with the magnitude of $\mathbf{F}(x, y)$. This vector must be tangent to any solution curve through (x, y). Simply said, the PPLANE6 Display window displays the phase plane for the planar system.

There are two buttons on the PPLANE6 Display window with the labels **Quit** and **Print**. There are several menus. Finally, below the vector field there is a small Cursor position window, and a larger message window through which `pplane` will communicate with the user. At this time it should contain the single word "Ready," indicating that it is ready to follow orders.

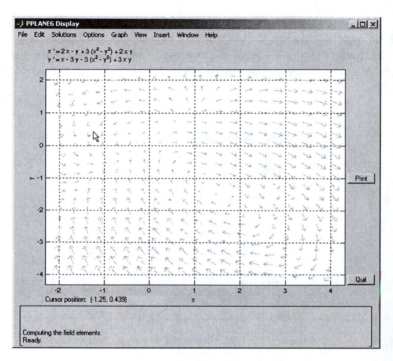

Figure 7.4. The display window for `pplane6`.

To compute and plot a solution curve from an initial point, move the mouse to that point, using the cursor position window to improve your accuracy, and click the mouse button. The solution will be computed and plotted, first in the direction in which the independent variable is increasing, and then in the opposite direction. After plotting several solutions, the display will look something like Figure 7.5. Notice that the solution curves all seem to start and stop in the same two points, and that the message window indicates that these are equilibrium points. We will have more to say about that later.

If you open the **Options** menu, you will see several ways of modifying how solutions are computed and plotted. The solution direction can be chosen by choosing **Options→Solution direction→Forward** or **Options→Solution direction→Back**. Some times it is desirable to indicate where the computation of a solution curve is started. This can be effected by selecting **Options→Mark initial points.**

In the **Edit** menu you will find the same "zoom in" and "zoom back" options available in `dfield6`.[2] Just as in `dfield6`, zooming can be done with the mouse.[3] For example, if you have a two button mouse, you can zoom at a point by clicking and and dragging with the right mouse button. There is an additional zoom option, **Zoom in square**, which we will describe in Chapter 13. You can print the PPLANE6 Display window to the default printer without the buttons and the message window by simply clicking the **Print** button. All of the print, save, and export options that are available for MATLAB figure windows are also available for the PPLANE6 Display window.

---

[2] See Chapter 3 for a review.
[3] See the inside cover of this manual for a summary of mouse button actions on various platforms.

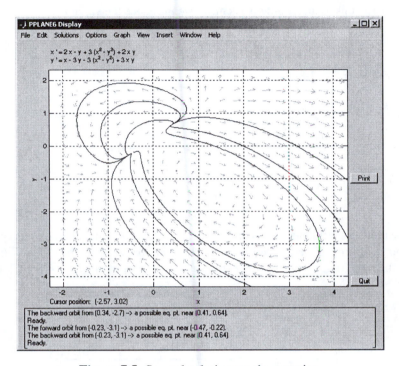

Figure 7.5. Several solutions to the equation.

Changing the System — Using the PPLANE6 Setup Window

We will illustrate the use of `pplane6` by using it to do a phase plane analysis of the motion of a pendulum. The differential equation for the motion of a pendulum is

$$mL\frac{d^2\theta}{dt^2} = -mg\sin(\theta) - c\frac{d\theta}{dt}, \tag{7.8}$$

where θ is the angular displacement of the pendulum from the vertical, L is the length of the pendulum arm, g is the acceleration due to gravity, and c is the damping constant. If we choose a convenient measure of time by setting $s = \sqrt{g/L}\, t$, the equation becomes[4]

$$\frac{d^2\theta}{ds^2} = -\sin(\theta) - a\frac{d\theta}{ds}, \tag{7.9}$$

where

$$a = \frac{c}{m\sqrt{gL}}$$

is again called the damping constant. Notice that this *scaling* of the variables reduces the number of parameters from four to one, and clearly indicates the importance of the *lumped* parameter a.

---

[4] Hint: $\frac{d\theta}{dt} = \frac{d\theta}{ds}\frac{ds}{dt} = \sqrt{\frac{g}{L}}\frac{d\theta}{ds}$; thus, $\frac{d^2\theta}{dt^2} = \frac{g}{L}\frac{d^2\theta}{ds^2}$ (why?).

97

We want to write this as a first order system, so we introduce the variables

$$x = \theta \quad \text{and} \quad y = \frac{d\theta}{ds}. \tag{7.10}$$

Then we have by (7.9) and (7.10)

$$\frac{dx}{ds} = \frac{d\theta}{ds} = y,$$

$$\frac{dy}{ds} = \frac{d^2\theta}{ds^2} = -\sin(\theta) - a\frac{d\theta}{ds} = -\sin(x) - ay, \tag{7.11}$$

or, more simply,

$$x' = y,$$
$$y' = -\sin(x) - ay, \tag{7.12}$$

where the prime indicates differentiation with respect to s. This is a planar autonomous system which we can analyze using pplane6. Enter the equations into the appropriate boxes the PPLANE6 Setup window This is exactly same as it is in dfield6, except that there are two equations. See Figure 7.6.

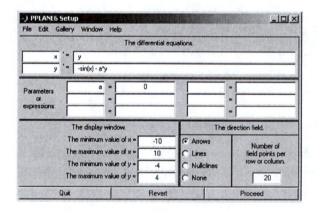

Figure 7.6. The PPLANE6 Setup window for the pendulum equation.

Since the damping constant a is not yet a number, we will assign it a value in one of the parameter boxes. For the time being, let's use the value $a = 0$. This value can be changed later to a positive number to see the phase plane for a damped pendulum. Next we have to describe the display rectangle. Since $x = \theta$ represents an angle, and we will want to plot a couple of full periods, we enter $-10 \leq x \leq 10$. For the dimensions in y we have to experiment. For now, $-4 \leq y \leq 4$ will do. Entering this data is just like it is in dfield6.

Finally, we decide which kind of direction field we want and how many field points we want displayed. These are available as menu options in dfield6. The choices are more important in pplane6, so they can be made directly in the PPLANE6 Setup window. Let's keep the default values. The completed PPLANE6 Setup window for the pendulum equation is shown in Figure 7.6. Click the **Proceed** button to transfer information from the PPLANE6 Setup window to the PPLANE6 Display window and start the computation of the direction field.

Plotting Solution Curves

Select **Options→Mark initial points** to mark the initial points of our solution curves. Select **Solutions→Keyboard input** and start a solution trajectory with initial condition $x(0) = 2$ and $y(0) = 0$. Note that this option behaves in a manner similar to that in dfield6. A closed orbit is plotted, giving the phase plane depiction of the standard motion of a pendulum shown in Figure 7.7. Go back to the PPLANE6 Setup window and change the parameter to $a = 0.5$, click proceed and then compute the solution as before. The result is shown in Figure 7.8.

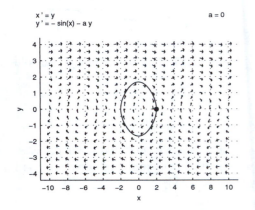

Figure 7.7. Phase-plane plot of the solution with initial condition $(x, y) = (2, 0)$ and $a = 0$.

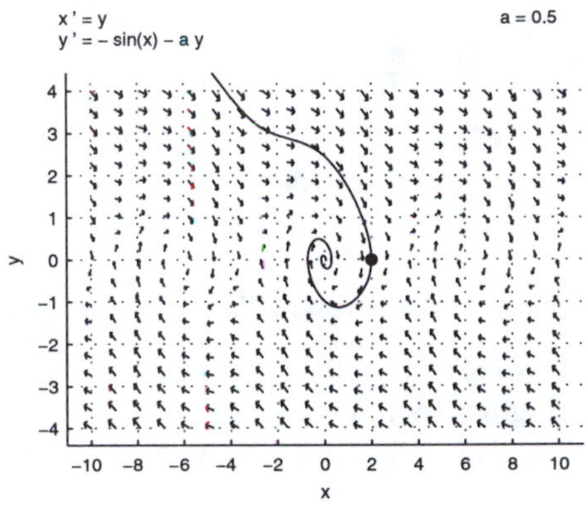

Figure 7.8. Phase-plane plot of the solution with initial condition $(x, y) = (2, 0)$ and $a = 0.5$.

Although the phase plane solutions pictured in Figures 7.7 and 7.8 are useful, we get a better understanding of a solution by looking at other graphical representations. This is easy to do in pplane6. Clicking on the **Graph** menu reveals five choices. Select **Graph→Both**. Notice that the vector field in the PPLANE6 Display window disappears and your cursor changes to a "cross hairs" as you move it over the figure. Center the cross hairs on the solution trajectory pictured in Figure 7.8 and click the (left) mouse button. This action should produce a plot in a new PPLANE6 t-plot window similar to that shown in Figure 7.9. Notice that both components of the solution are plotted against the independent variable t. In addition, this figure contains a legend identifying the two curves, and radio buttons that enable easy selection of five different views of the solution. These are the same options that appear in the **Graph** menu.

The plot in Figure 7.9 would be more interesting if the independent variable were restricted to lie between -2 and 10. We can accomplish this by cropping the figure. Position the cursor over the figure at $t = -2$. Then click the (left) mouse button, drag to $t = 10$, and release the mouse button. Notice that a green line appears to help you in this process. The **Crop** button is now enabled, and when you click it a new t-plot figure opens with the t variable limited to $-2 \leq t \leq 10$.

The most interesting of the plot options is the Composite plot. Click on that radio button, and you get Figure 7.10. This is basically a 3-dimensional plot of both components of the solution versus t. However, it also contains the other views as projections onto the three coordinate planes. The composite

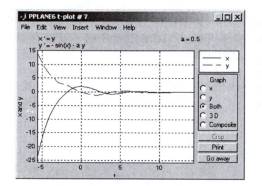

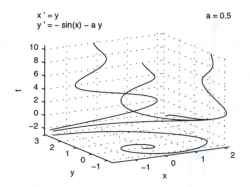

Figure 7.9. Plot of x and y versus t for the solution in Figure 7.8.

Figure 7.10. Composite plot of the solution.

plot lets you understand how the various graphical representations are related.

The composite and 3D views can be rotated so you can see them from different perspectives. Select **View→Figure Toolbar** to bring up the toolbar. Then select the rotate tool, the icon on the far right of the toolbar. Now click and drag in the figure window to rotate the view.

Other Properties of PPLANE6

Editing windows. The PPLANE6 Display window provides several editing commands that are similar to those in the DFIELD6 Display window. In fact there are a few more because of the additional features in `pplane6`. In particular, you can delete any graphics object, including text, by selecting **Edit→Delete a graphics object** and clicking on the object you wish to delete.

The PPLANE6 Display and t-plot figure windows can be edited just like any MATLAB figure window. However, use the property editors with caution, since they interfere with the interactive features of `pplane6`. The use of the rotate feature is an exception to this rule.

Printing, saving, and using the clipboard. You can print or export the `pplane6` windows in the standard ways described in Chapter2. The **Print** buttons on the PPLANE6 Display and t-plot windows will cause the figure to be printed to the default printer without the various buttons.

Saving and loading PPLANE6 systems and galleries. Again, as is the case for `dfield6`, the information on the setup window can be added to the **Gallery** menu, and can be saved in files. The files containing information about systems have the suffix `.pps`. You can also make, save, and load entirely new galleries that contain a number of important systems for a lecture or presentation. Gallery files have the suffix `.ppg`. The mechanisms for working with systems and galleries are the same as explained in Chapter 3 for `dfield6`.

Exercises

The planar, autonomous system having the form

$$x' = ax + by,$$
$$y' = cx + dy,$$

where a, b, c, and d are arbitrary real numbers, is called a *linear* system of first order differential equations. Exercises 1 – 6 each contain a solution of some linear system. Use MATLAB to create a plot of x versus t, y versus t, and a plot of y versus x in the phase plane. Use the subplot command, as in

```
subplot(221),plot(t,x),axis tight
subplot(222),plot(t,y),axis tight
subplot(212),plot(x,y),axis equal
```

to produce a plot containing all three plots. Use the suggested time interval.

1. $x = -2e^{-2t} + 3e^{-3t}$
   $y = 4e^{-2t} - 3e^{-3t}$
   $[-0.5, 2]$

2. $x = -4e^{2t} + 3e^{3t}$
   $y = -2e^{2t} + 3e^{3t}$
   $[-2, 0.5]$

3. $x = -2e^{-2t} - e^{3t}$
   $y = e^{-2t} + e^{3t}$
   $[-1.5, 1.0]$

4. $x = \cos 2t - 3 \sin 2t$
   $y = 2 \sin 2t + \cos 2t$
   $[-2\pi, 4\pi]$

5. $x = e^{t}(\cos t + 5 \sin t)$
   $y = e^{t}(-8 \sin t + \cos t)$
   $[-\pi, \pi/2]$

6. $x = e^{-t}(\cos t + \sin t)$
   $y = e^{-t}(\cos t - \sin t)$
   $[-\pi, \pi]$

For Exercises 7 – 12, select **Gallery→linear system** in the PPLANE6 Setup window. Adjust the parameters to match the indicated linear system. Accept all other default settings in the PPLANE6 Setup window. Select **Solutions→ Keyboard input** in the PPLANE6 Display window to start a solution trajectory with the given initial condition. Finally, select **Graph→Both** and click your solution trajectory to obtain plots of x and y versus t. For Exercise n, compare your output to Exercise $n - 6$ above.

7. $x' = -4x - y$
   $y' = 2x - y$
   $x(0) = 1, y(0) = 1$

8. $x' = x + 2y$
   $y' = -x + 4y$
   $x(0) = -1, y(0) = 1$

9. $x' = -7x - 10y$
   $y' = 5x + 8y$
   $x(0) = -3, y(0) = 2$

10. $x' = -2x - 4y$
    $y' = 2x + 2y$
    $x(0) = 1, y(0) = 1$

11. $x' = 4x + 2y$
    $y' = -5x - 2y$
    $x(0) = 1, y(0) = 1$

12. $x' = -x + y$
    $y' = -x - y$
    $x(0) = 1, y(0) = 1$

13. In the predator-prey system

$$L' = -L + LP,$$
$$P' = P - LP,$$

L represents a lady bug population and P represents a pest that lady bugs like to eat. Enter the system in the PPLANE6 Setup window, set the display window so that $0 \leq L \leq 2$ and $0 \leq P \leq 3$, then select **Arrows** for the direction field.

a) Use the Keyboard input window to start a solution trajectory with initial condition $(0.5, 1.0)$. Note that the lady bug-pest population is periodic. As the lady bug population grows, their food supply dwindles and the lady bug population begins to decay. Of course, this gives the pest population time to flourish, and the resulting increase in the food supply means that the lady bug population begins to grow. Pretty soon, the populations come full cycle to their initial starting position.

b) Suppose that the pest population is harmful to a farmer's crop and he decides to use a poison spray to reduce the pest population. Of course, this will also kill the lady bugs. Question: Is this a wise idea? Adjust the system as follows:

$$L' = -L + LP - HL,$$
$$P' = P - LP - HP.$$

Note that this model assumes that the growth rate of each population is reduced by a fixed percentage of the population. Enter this system in the PPLANE6 Setup window, but keep the original display window settings. Create and set a parameter H = 0.2. Start a solution trajectory with initial condition (0.5, 1.0).

c) Repeat part b) with $H = 0.4$, 0.6, and 0.8. Is this an effective way to control the pests? Why? Describe what happens to each population for each value of the parameter H.

14. Enter the system

$$x' = -\cos y + 2y \cos y^2 \cos 2x$$
$$y' = -\sin x + 2 \sin y^2 \sin 2x$$

in the PPLANE6 Setup window. Set the display window to $-10 \le x \le 10$ and $-2 \le y \le 4$ and select **None** for the vector field.

a) Select **File→Save the current system** and save the system with the name teddybears.pps.

b) Select **Gallery→linear system** to load the linear system template. Select **File→Load a system** and load the system teddybears.pps.

c) Select **Solutions→Keyboard input** and start a solution trajectory with initial condition $(\pi/2, 0)$ to create the "legs" of the Teddy bears, and a trajectory with initial condition $(-7.2, 3.5)$ to create the "heads" of the Teddy bears.

d) Experiment further with this "wild and wooly" example. Click the mouse in the phase plane to start other solution trajectories. Can you find the "eyes" of the Teddy bears?

When you portray sufficient solution trajectories in the phase plane so as to determine all of the important behavior of a planar, autonomous system, you have created what is called a "phase portrait." In Exercises 15 – 22, use pplane6 to create a phase portrait for the indicated system on the prescribed display window. Take special notice of where solution curves end, as reported in the message window.

15. $x' = y$
$y' = (1 - y^2)y - x$
$-3 \le x \le 3, \ -3 \le y \le 3$

16. $x' = y + (y^2 - x^2 + 0.5x^4)(1 - x^2)$
$y' = x(1 - x^2) - y(y^2 - x^2 + 0.5x^4)$
$-3 \le x \le 3, \ -2 \le y \le 2$

17. $x' = x(2y^3 - x^3)$
$y' = -y(2x^3 - y^3)$
$-3 \le x \le 3, \ -3 \le y \le 3$

18. $x' = y + x^2 + xy$
$y' = -x + xy + y^2$
$-2 \le x \le 2, \ -2 \le y \le 2$

19. $x' = -y - x(x^2 - y^2)$
$y' = -x - y(x^2 - y^2))$
$-3 \le x \le 3, \ -3 \le y \le 3$

20. $x' = y - x(x^2 + y^2 - 1)$
$y' = -x - y(x^2 + y^2 - 1)$
$-2 \le x \le 2, \ -2 \le y \le 2$

21. $x' = x^2 - y^2$
$y' = 2xy$
$-1 \le x \le 1, \ - \le y \le 1$

22. $x' = x^3 - 3xy^2$
$y' = 3xy^2 - y^3$
$-1 \le x \le 1, \ -1 \le y \le 1$

The equation $my'' + dy' + ky = 0$ represents a damped, spring-mass system. In Exercises 23 – 26, values of the parameters m, d, and k have been chosen to create a specific example of a damped, spring-mass system. We are looking for the solution y with the given initial position and velocity. Let

$$x_1 = y, \quad \text{and} \quad x_2 = y',$$

and use this change of variables to write each spring-mass system as a planar, autonomous system. Use pplane6 to obtain a printout of the graph of y versus t.

23. $y'' + 2y' + 2y = 0$
    $y(0) = 0,\ y'(0) = -1$

24. $y'' + 4y' + 4y = 0$
    $y(0) = -2,\ y'(0) = 2$

25. $y'' + 3y' + 2y = 0$
    $y(0) = 3,\ y'(0) = 2$

26. $y'' + 4y = 0$
    $y(0) = -4,\ y'(0) = -2$

27. Tanks A and B contain mixtures of water and a pollutant. Tank A has a 100 gallon capacity and is full. Tank B is also full and has a 200 gallon capacity. Initially, tank A contains 40 pounds of pollutant and tank B contains pure water (no pollutant). At $t = 0$, pure water begins pouring into tank A at a rate of 10 gallons per minute. The mixture flows from tank A into tank B at a rate of 10 gallons per minute, and drains out of tank B at 10 gallons per minute. If x_A and x_B represent the number of pounds of pollutant in tank A and tank B, respectively, show that

$$x'_A = -\frac{1}{10}x_A,$$

$$x'_B = \frac{1}{10}x_A - \frac{1}{20}x_B,$$

where $x_A(0) = 40$ and $x_B(0) = 0$.

   a) Use `pplane6` and its zoom tools to estimate the maximum amount of pollutant in tank B and the time that it occurs.

   b) Use `pplane6` to estimate the time that it takes the pollutant in tank B to reach a level of 10 pounds. *Note:* This event occurs at two separate times, once as the pollutant level is rising in tank B, and a second time when the pollutant level is decreasing.

28. Fasten a length of string to a hook. Securely fasten a mass to the the other end of the string. Let θ represent the displacement of the mass from the vertical. Assume that θ is measured in radians, with positive displacements in the counterclockwise direction. Let ω represent the angular velocity of the mass, with positive angular velocity in the counterclockwise direction. Displace the mass 30° counter-clockwise (positive $\pi/6$ radians) from the vertical and release the mass from rest. Let the pendulum decay to a stable equilibrium point at $\theta = 0,\ \omega = 0$.

   Important Note. You will not benefit as much as you should from this exercise if you don't first complete parts a) and b) before attempting part c).

   a) Without using any technology, sketch graphs of θ versus t and ω versus t approximating the motion of the pendulum.

   b) Without using any technology, sketch graphs of ω versus θ. Place ω on the vertical axis, θ on the horizontal axis. *Note:* This is a lot harder than it looks. We suggest that you work with a partner or a group and compare solutions before moving on to part c).

   c) Select **Gallery→pendulum** in the PPLANE6 Setup window. Adjust the damping parameter to D=0.1, and set the display window so that $-1 \le \theta \le 1$ and $-1 \le \omega \le 1$. Select **Options→Solution direction→Forward** and use the Keyboard input window to start a solution trajectory with initial condition $\theta(0) = \pi/6$ and $\omega(0) = 0$. Compare this result with your hand-drawn solution in part b). Select **Graph→Both** and click your solution trajectory in the phase plane to produce plots of θ versus t and ω versus t. Compare these with your hand-drawn solutions in part a).

29. Repeat the pendulum experiment of the previous problem. Displace the mass 30° counter-clockwise (positive $\pi/6$ radians) from the vertical, only this time do not release the mass from rest. Instead, push the mass in the clockwise (negative) direction with enough negative angular velocity so that it spins around in a circle exactly one time before settling into a motion decaying to a stable equilibrium. The tricky part of this experiment is the fact that the stable equilibrium point is now $\theta = -2\pi,\ \omega = 0$.

   a) Without using any technology, sketch graphs of θ versus t and ω versus t approximating the motion of the pendulum.

   b) Without using any technology, sketch graphs of ω versus θ. Place ω on the vertical axis, θ on the horizontal axis. *Note:* This is a lot harder than it looks. We suggest that you work with a partner or a group and compare solutions before moving on to part c).

   c) Select **Gallery→pendulum** in the PPLANE6 Setup window. Adjust the damping parameter to D=0.1, and set the display window so that $-10 \le \theta \le 5$ and $-4 \le \omega \le 4$. Select **Options→Solution**

direction→**Forward** and use the Keyboard input window to start a solution trajectory with initial condition $\theta(0) = \pi/6$ rad and $\omega(0) = -2.5$ rad/s. Compare this result with your hand-drawn solution in part b). Select **Graph**→**Both** and click your solution trajectory in the phase plane to produce plots of θ versus t and ω versus t. Compare these with your hand-drawn solutions in part a).

30. Consider the predator-prey system

$$R' = R - RF,$$
$$F' = -F + RF,$$

where R and F represent rabbit and fox populations, respectively. Enter the system in `pplane6` and set the display window so that $0 \le R \le 2$ and $0 \le F \le 2$. Use Keyboard input to start a solution trajectory at $R = 0.5$ and $F = 1$. Note that the trajectory is periodic. Use `pplane6` to estimate the time it takes to travel this periodic trajectory exactly once; i.e., find the period of the oscillation. *Hint:* Try cropping an x versus t plot.

31. Start a new session by exiting and restarting `pplane6`. Select **File**→**Delete the current gallery** in the PPLANE6 Setup window. Examine your textbook and select three planar, autonomous systems of interest. Enter the first system and set the parameters (if any), then adjust the display window and select the direction field. Select **Gallery**→**Add current system to the gallery** to add your system to the gallery. Add each of your remaining systems to the gallery, then select **File**→**Save a gallery** to create and save your new gallery.

    a) Start a new session by exiting and restarting `pplane6`. Delete the current gallery, then load your newly created gallery.

    b) Check that each system in the gallery works as it should.

8.   Solving ODEs in MATLAB

This is one of the most important chapters in this manual. Here we will describe how to use MATLAB's built-in numerical solvers to find approximate solutions to almost any system of differential equations. At this point readers might be thinking, "I've got `dfield6` and `pplane6` and I'm all set. I don't need to know anything further about numerical solvers." Unfortunately, this would be far from the truth. For example, how would we handle a system modeling a driven, damped oscillator such as

$$
\begin{aligned}
x_1' &= x_2, \\
x_2' &= -2x_1 - 3x_2 + \cos t.
\end{aligned}
\tag{8.1}
$$

System (8.1) is **not autonomous**! Pplane6 can only handle autonomous systems. Furthermore, how would we handle a system such as

$$
\begin{aligned}
x_1' &= x_1 + 2x_2, \\
x_2' &= -x_3 - x_4, \\
x_3' &= x_1 + x_3 + x_4, \\
x_4' &= -2x_2 - x_4.
\end{aligned}
\tag{8.2}
$$

The right-hand sides of system (8.2) do not explicitly involve the independent variable t, so the system is autonomous. However, system (8.2) is **not planar**! Pplane6 can only handle autonomous systems of two first order differential equations (planar systems).

Finally, in addition to handling systems of differential equations that are neither planar nor autonomous, we have to develop some sense of trust in the output of numerical solvers. How accurate are they?

MATLAB's ODE Suite

A new suite of ODE solvers was introduced with version 5 of MATLAB. The suite now contains the seven solvers `ode23`, `ode45`, `ode113`, `ode15s`, `ode23s`, `ode23t`, and `ode23tb`. We will spend most of our time discussing the general purpose solver `ode45`. We will also briefly discuss `ode15s`, an excellent solver for stiff systems. Although we will not discuss the other solvers, it is important to realize that the calling syntax is the same for each solver in MATLAB's suite. In the next chapter, we will discuss the MATLAB function `odesolve`, which puts a graphical user interface around the MATLAB solvers.

In addition to being very powerful, MATLAB's numerical solvers are easy to use, as you will soon discover. The methods described herein are used regularly by engineers and scientists, and are available in any version of MATLAB. The student who learns the techniques described here will find them useful in many later circumstances.

The `ode45` solver uses a variable step Runge-Kutta procedure. Six derivative evaluations are used to calculate an approximation of order five, and then another of order four. These are compared to come up with an estimate of the error being made at the current step. It is required that this estimated error at any step should be less than a predetermined amount. This predetermined amount can be changed using two tolerance parameters, and a very high degree of accuracy can be required of the solver. We will discuss this later in this chapter. For most of the uses in this Manual the default values of the tolerance parameters will provide sufficient accuracy, and we will first discuss the use of the solver with its default settings.

Single First Order Differential Equations

We are looking at an initial value problem of the form $x' = f(t, x)$, with $x(t_0) = x_0$. The calling syntax for using `ode45` to find an approximate solution is `ode45(odefcn,tspan,x0)`, where `odefcn` calls for a functional evaluation of $f(t, x)$, `tspan=[t0,tfinal]` is a vector containing the initial and final times, and `x0` is the x-value of the initial condition.

Example 1. *Use* `ode45` *to plot the solution of the initial value problem*

$$x' = \frac{\cos t}{2x - 2}, \quad x(0) = 3, \tag{8.3}$$

on the interval $[0, 2\pi]$.

Note that equation (8.3) is in normal form $x' = f(t, x)$, where $f(t, x) = \cos t/(2x - 2)$. From the data we have `tspan = [0,2*pi]` and `x0 = 3`. We need to encode the `odefcn`. Open your editor and create an ODE function M-file with the contents

```
function xprime = ch8examp1(t,x)
xprime = cos(t)/(2*x - 2);
```

Save the file as `ch8examp1.m`. Notice that $f(0, 2) = \cos 0/(2 \cdot 2 - 2) = 1/2$ and

```
>> ch8examp1(0,2)
ans =
    0.5000
```

so the M-file gives the correct answer. If your output does not match ours, check your function code for errors. If you still have difficulty, revisit Chapter 4, especially the discussion of the MATLAB path.

You can get a fast plot of the solution by executing `ode45(@ch8examp1,[0,2*pi],3)`.[1] Notice that we did not specify any outputs. This causes `ode45` to plot the solution. Usually you will want the output in order to analyze it and treat it in your own way. The following code should produce an image similar to that shown in Figure 8.1.

```
>> [t,x] = ode45(@ch8examp1,[0,2*pi],3);
>> plot(t,x)
>> title('The solution of x''=cos(t)/(2x - 2), with x(0) = 3.')
>> xlabel('t'), ylabel('x'), grid
```

The output of the `ode45` command consists of a column vector `t`, containing the t-values at which the solution was computed, and a column vector `x` containing the computed x-values.[2]

---

[1] If you are using a version of MATLAB prior to version 6.0, the calling sequence should be `ode45('ch8examp1',[0,2*pi],3)`. We discussed this issue in Chapter 4. There are other differences in the syntax to be used with MATLAB's ODE solvers between version 6 and prior versions. We will emphasize the version 6 usage. For systems without parameters, users of older versions of MATLAB need only replace the function handles by the function names between single quotes. As further differences arise, we will explain them, but if you are using an older version, you should execute `help ode45` to be sure of the correct syntax.

[2] To see the output from the `ode45` routine execute `[t,x]`.

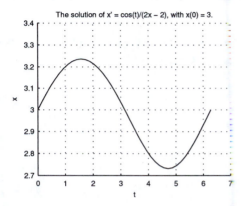

Figure 8.1. The solution of the initial value problem in Example 1.

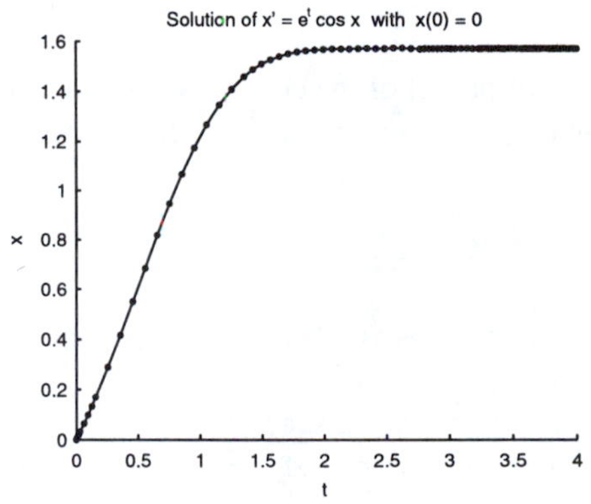

Figure 8.2. The solution of the initial value problem in Example 2.

Using inline functions. If you do not want to save your work, the easiest way to encode the needed odefcn is to create the inline function

```
>> f = inline('cos(t)/(2*x - 2)','t','x')
f =
     Inline function:
     f(t,x) = cos(t)/(2*x - 2)
```

We notice that $f(0, 2) = 1/2$, and

```
>> f(0,2)
ans =
    0.5000
```

so it seems to be working correctly. The command ode45(f,[0,2*pi],3) [3] will plot the solution in a figure window, with the computed points plotted as circles. Try it yourself. If you want to save the computed data, execute [t,x] = ode45(f,[0,2*pi],3);.

Function M-File Drivers and Subfunctions[4]

In Chapter 4 we explained how to use function M-files to ease the process of doing computations and creating reproducible graphics. This technique is especially useful when using ODE solvers, as the next example will illustrate.

Example 2. *Use* ode45 *to plot the solution of the initial value problem*

$$\frac{dx}{dt} = e^t \cos x, \quad x(0) = 0, \tag{8.4}$$

---

[3] Notice that we use just f when we are using an inline function. Function handles and single quotes are not needed.

[4] Subfunctions were introduced in version 6 of MATLAB and are not available in earlier versions.

on the interval [0, 4]. *Plot the computed points in a distinctive marker style as well as the curve connecting them.*

We will put all of the commands into a function M-file, with the `odefcn` as a subfunction. We will also plot the computed points with dots to make them visible. Create the file

```
function ch8examp2
close all
[t,x] = ode45(@dfile,[0,4],0);
plot(t,x,'.-')
xlabel('t'), ylabel('x')
title('Solution of x'' = e^t cos x   with   x(0) = 0')

function xprime = dfile(t,x)
xprime = exp(t)*cos(x);
```

and save it as `ch8examp2.m`. Executing `ch8examp2` will create Figure 8.2. Notice in Figure 8.2 that the computed points start out close to each other and then spread out for a while before becoming really close to each other as t gets larger than 3. This illustrates the fact that `ode45` is a variable step solver.

The file `ch8examp2` is a function M-file which serves as a master routine or driver, with the ODE function defined as a subfunction. Subfunctions are available only within the function where they are defined. Notice that `ch8examp2` cannot be a script M-file, since subfunctions are allowed only in function M-files.

There are several advantages to using subfunctions inside master functions in this way. Once you have written the master file for a particular purpose, it is easy to reproduce the result. It puts everything involved with a problem in one file. This file can be given a suggestive name, as we did here, to make it easy to interpret later. With all of the commands in one file, it is relatively easy to make changes and correct errors. Furthermore, once you have written one such file, additional ones might well be simple modifications of the first. Frequently, all that is needed is to change the values of the various properties. This applies to the ODE subfunction as well. There is really no need to give this subfunction a distinctive name. Call it `dfile`, or just `f`. Since the subfunction is available only within the master function it does not matter if it has the same name in every such file. Finally, notice that it is not really more work to produce the file than to to execute the original commands.

Systems of First Order Equations

Actually, systems are no harder to handle using `ode45` than are single equations. Consider the following system of n first order differential equations:

$$
\begin{aligned}
x_1' &= f_1(t, x_1, x_2, \ldots, x_n), \\
x_2' &= f_2(t, x_1, x_2, \ldots, x_n), \\
&\vdots \\
x_n' &= f_n(t, x_1, x_2, \ldots, x_n).
\end{aligned}
\tag{8.5}
$$

108

System (8.5) can be written in the vector form

$$\begin{bmatrix} x_1 \\ x_2 \\ \vdots \\ x_n \end{bmatrix}' = \begin{bmatrix} x_1' \\ x_2' \\ \vdots \\ x_n' \end{bmatrix} = \begin{bmatrix} f_1(t, [x_1, x_2, \ldots, x_n]^T) \\ f_2(t, [x_1, x_2, \ldots, x_n]^T) \\ \vdots \\ f_n(t, [x_1, x_2, \ldots, x_n]^T) \end{bmatrix} \qquad (8.6)$$

If we define the vector $\mathbf{x} = [x_1, x_2, \ldots, x_n]^T$, system (8.6) becomes

$$\mathbf{x}' = \begin{bmatrix} f_1(t, \mathbf{x}) \\ f_2(t, \mathbf{x}) \\ \vdots \\ f_n(t, \mathbf{x}) \end{bmatrix}. \qquad (8.7)$$

Finally, if we define $\mathbf{F}(t, \mathbf{x}) = [f_1(t, \mathbf{x}), f_2(t, \mathbf{x}), \ldots, f_n(t, \mathbf{x})]^T$, then system (8.7) can be written as

$$\mathbf{x}' = \mathbf{F}(t, \mathbf{x}), \qquad (8.8)$$

which, most importantly, has a form identical to the single first order differential equation, $x' = f(t, x)$, used in Example 1. Consequently, if `function xprime=F(t,x)` is the first line of a function ODE file, it is extremely important that you understand that x is a vector[5] with entries `x(1)`, `x(2)`,..., `x(n)`. Confused? Perhaps a few examples will clear the waters.

Example 3. *Use* `ode45` *to solve the initial value problem*

$$\begin{aligned} x_1' &= x_2 - x_1^2, \\ x_2' &= -x_1 - 2x_1 x_2, \end{aligned} \qquad (8.9)$$

on the interval [0, 10], *with initial conditions* $x_1(0) = 0$ *and* $x_2(0) = 1$.

We can write system (8.9) as the vector equation

$$\begin{bmatrix} x_1 \\ x_2 \end{bmatrix}' = \begin{bmatrix} x_2 - x_1^2 \\ -x_1 - 2x_1 x_2 \end{bmatrix}. \qquad (8.10)$$

If we set $\mathbf{F}(t, [x_1, x_2]^T) = \left[x_2 - x_1^2, -x_1 - 2x_1 x_2 \right]^T$ and $\mathbf{x} = [x_1, x_2]^T$, then system (8.10) takes the form $\mathbf{x}' = \mathbf{F}(t, \mathbf{x})$. The key thing to realize is the fact that $\mathbf{x} = [x_1, x_2]^T$ is a vector with two components, x_1 and x_2. Similarly, $\mathbf{x}' = [x_1', x_2']^T$ is also a vector with two components, x_1' and x_2'.

Open your editor and create the following ODE file[6].

```
function xprime = F(t,x)
xprime = zeros(2,1);     %The output must be a column vector
xprime(1) = x(2) - x(1)^2;
xprime(2) = -x(1) - 2*x(1)*x(2);
```

---

[5] In MATLAB, if `x=[2;4;6;8]`, then `x(1)=2`, `x(2)=4`, `x(3)=6`, and `x(4)=8`.

[6] Even though this system is autonomous, the ODE suite does not permit you to write the first line of the ODE file without the t variable as in `function xprime=F(x)`. If you did, you would receive lots of error messages.

Save the file as F.m. The line xprime = zeros(2,1) *initializes* xprime, creating a *column* vector with two rows and 1 column. Each of its entries is a zero. Note that the next two lines of the ODE file duplicate exactly what you see in system (8.9).

Even though we write $\mathbf{x} = [x_1, x_2]^T$ in the narrative, it is extremely important that you understand that $\mathbf{x}$ is a *column* vector; that is,

$$\mathbf{x} = \begin{bmatrix} x_1 \\ x_2 \end{bmatrix}.$$

Therefore, the variable x in the ODE file is also a *column* vector.

As always, it is an excellent idea to test that your function ODE file is working properly before continuing. Because $\mathbf{F}(t, [x_1, x_2]^T) = [x_2 - x_1^2, -x_1 - 2x_1x_2]^T$, we have that $\mathbf{F}(2, [3, 4]^T) = [4 - 3^2, -3 - 2(3)(4)]^T = [-5, -27]^T$. Test this result at the MATLAB prompt.

```
>> F(2,[3;4])
ans   =
     -5
    -27
```

Notice that the function F(t,x) expects two inputs, a scalar t and a column vector x. We used t = 2 and x = [3;4]. Because this system is autonomous and independent of t, the command F(5,[3;4]) should also produce [-5;-27]. Try it!

Before we call ode45, we must establish the initial condition. Recall that the initial conditions for system (8.9) were given as $x_1(0) = 0$ and $x_2(0) = 1$. Consequently, our initial condition vector will be

$$\mathbf{x}(0) = \begin{bmatrix} x_1(0) \\ x_2(0) \end{bmatrix} = \begin{bmatrix} 0 \\ 1 \end{bmatrix}.$$

Of course, this is a column vector, so we use x0 = [0;1] or x0 = [0, 1]'. Then

```
>> [t,x] = ode45(@F,[0,10],[0;1]);
```

computes the solution.[7]

Typing whos at the MATLAB prompt yields[8]

```
>> whos
  Name        Size                       Bytes  Class

  t           77x1                          616  double array
  x           77x2                         1232  double array
```

Note that t and x both have 77 rows. Each row in the solution array x corresponds to a time in the same row of the column vector t. The matrix x has two columns. The first column of x contains the solution

---

[7] If you are working with version 5 of MATLAB, execute [t,x]=ode45('F',[0,10],[0;1]) instead.

[8] You may have more variables in your workspace than the number shown here. In addition the number of rows in t and x may differ from system to system.

for x_1, while the second column of x contains the solution x_2. Typing [t,x] at the MATLAB prompt and viewing the resulting output will help make this connection[9].

Given a column vector v and a matrix A which have the same number of rows, the MATLAB command plot(v,A) will produce a plot of each column of the matrix A versus the vector v.[10] Consequently, the command plot(t,x) should produce a plot of each column of the matrix x versus the vector t. The following commands were used to produce the plots of x_1 versus t and x_2 versus t in Figure 8.3.

```
>> plot(t,x)
>> title('x_1'' = x_2 - x_1^2 and x_2'' = -x_1 - 2x_1x_2')
>> xlabel('t'), ylabel('x_1 and x_2')
>> legend('x_1','x_2'), grid
```

Since we are limited to black and white in this Manual, the different colors of the curves are not available, so we changed the curve for x_2 to a dotted linestyle. To do this we clicked the mouse button near the x_2 curve and selected **Edit→Current Object Properties ....** We then made the change of line style in the Editor.

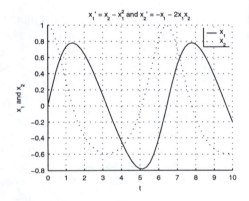

Figure 8.3. plot(t,x) plots both components of the solution.

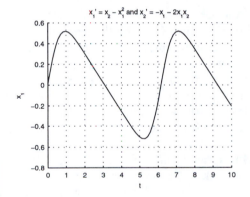

Figure 8.4. plot(t,x(:,1)) plots the first component of the solution.

If only the first component of the solution is wanted, enter plot(t,x(:,1)). The colon in the notation x(:,1) indicates[11] that we want all rows, and the 1 indicates that we want the first column. The result is shown in Figure 8.4. Similarly, if only the second component is wanted, enter plot(t,x(:,2)). This is an example of the very sophisticated indexing options available in MATLAB. It is also possible to plot the components of the solution against each other with the commands

```
>> plot(x(:,1),x(:,2))
>> title('x_1'' = x_2-x_1^2 and x_2'' = -x_1-2x_1x_2')
>> xlabel('x_1'), ylabel('x_2'), grid
```

---

[9] For example, note that the time values in the vector t run from 0 through 10, as they should.

[10] Execute help plot to see a complete description of what the plot command is capable of.

[11] Some MATLAB users pronounce the notation x(:,1) as "x, all rows, first column."

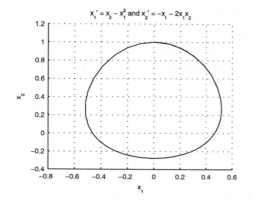

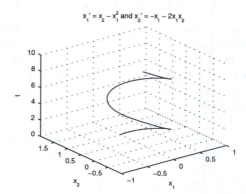

Figure 8.5. Phase plane plot of the solution.

Figure 8.6. 3D plot of the solution.

The result is called a *phase plane* plot, and it is shown in Figure 8.5.

Another way to present the solution to the system graphically is in a three dimensional plot, where both components of the solution are plotted as separate variables against the independent variable t. MATLAB does this using the command `plot3`. For example, enter `plot3(t,x(:,1),x(:,2))` to see the plot with t along the x-axis, and the two components of the solution along the y-axis and z-axis, respectively. Alternatively, enter `plot3(x(:,1),x(:,2),t)` to see the solution with t along the z-axis, and the two components of the solution along the x-axis and y-axis, respectively. The result of this command is shown in Figure 8.6. The three dimensional graph can be rotated by first selecting the rotation icon in the toolbar, and then clicking and dragging on the figure.

Solving systems with eul, rk2, and rk4. These solvers, introduced in Chapter 5, can also be used to solve systems. The syntax for doing so is not too different from that used with ode45. For example, to solve the system in (8.9) with `eul`, we use the command

```
>> [t,x] = eul(@F,[0,10],[0;1],h);
```

where h is the chosen step size. To use a different solver it is only necessary to replace `eul` with the `rk2` or `rk4`. Thus, the only difference between using one of these solvers and ode45 is that it is necessary to add the step size as an additional parameter.

Second Order Differential Equations

To solve a single second order differential equation it is necessary to replace it with the equivalent first order system. For the equation

$$y'' = f(t, y, y'), \tag{8.11}$$

we set $x_1 = y$, and $x_2 = y'$. Then $\mathbf{x} = [x_1, x_2]^T$ is a solution to the first order system

$$
\begin{aligned}
x_1' &= x_2, \\
x_2' &= f(t, x_1, x_2).
\end{aligned}
\tag{8.12}
$$

112

Conversely, if $\mathbf{x} = [x_1, x_2]^T$ is a solution of the system in (8.12), we set $y = x_1$. Then we have $y' = x_1' = x_2$, and $y'' = x_2' = f(t, x_1, x_2) = f(t, y, y')$. Hence y is a solution of the equation in (8.11).

Example 4. *Plot the solution of the initial value problem*

$$y'' + yy' + y = 0, \quad y(0) = 0, \ y'(0) = 1, \tag{8.13}$$

on the interval $[0, 10]$.

First, solve the equation in (8.13) for y''.

$$y'' = -yy' - y \tag{8.14}$$

Introduce new variables for y and y'.

$$x_1 = y \quad \text{and} \quad x_2 = y' \tag{8.15}$$

Then we have by (8.14) and (8.15),

$$x_1' = y' = x_2, \quad \text{and} \quad x_2' = y'' = -yy' - y = -x_1x_2 - x_1,$$

or, more simply,

$$\begin{aligned} x_1' &= x_2, \\ x_2' &= -x_1x_2 - x_1. \end{aligned} \tag{8.16}$$

If we let

$$\mathbf{F}(t, [x_1, x_2]^T) = \begin{bmatrix} x_2 \\ -x_1x_2 - x_1 \end{bmatrix} \tag{8.17}$$

and $\mathbf{x} = [x_1, x_2]^T$, then $\mathbf{x}' = [x_1', x_2']^T$ and system (8.16) takes the form $\mathbf{x}' = \mathbf{F}(t, \mathbf{x})$.

Let's address this example by putting everything into one master file. This file could be

```
function ch8examp4
[t,x] = ode45(@dfile,[0,10],[0;1]);
plot(t,x(:,1))
title('y'''' + yy'' + y = 0, y(0) = 0, y''(0) = 1')
xlabel('t'), ylabel('y'), grid

function xprime = dfile(t,x)
xprime = zeros(2,1);
xprime(1) = x(2);
xprime(2) = -x(1)*x(2) - x(1);
```

Save the file as ch8examp4.m. When we execute ch8examp4 we get Figure 8.7.

Let's examine the file ch8examp4. First notice that the subfunction dfile is the ODE file, and implements the function $\mathbf{F}$ defined in (8.17). According to (8.13), the initial condition is

$$\mathbf{x}(0) = \begin{bmatrix} x_1(0) \\ x_2(0) \end{bmatrix} = \begin{bmatrix} y(0) \\ y'(0) \end{bmatrix} = \begin{bmatrix} 0 \\ 1 \end{bmatrix}.$$

Hence the second line in ch8examp4 calls ode45 on the equation defined in the subfunction with the proper initial conditions. It is important to note that the original question called for a solution of $y'' + yy' + y = 0$. Recall from (8.15) that $y = x_1$. Consequently, the plot command in ch8examp4 will plot y versus t, as shown in Figure 8.7. If we replace the plot command with plot(x(:,1),x(:,2)) we will get a phase-plane plot like that in Figure 8.8.

113

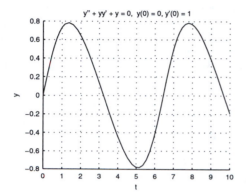

Figure 8.7. The solution to the initial value problem (8.13).

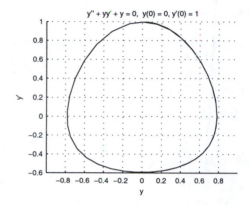

Figure 8.8. Phase-plane plot of the solution.

The Lorenz System and Passing Parameters

The solvers in MATLAB can solve first order systems containing as many equations as you like. As an example we will solve the Lorenz system. This is a system of three equations published in 1963 by the meteorologist and mathematician E. N. Lorenz. It represents a simplified model for atmospheric turbulence beneath a thunderhead. The equations are

$$
\begin{aligned}
x' &= -ax + ay, \\
y' &= rx - y - xz, \\
z' &= -bz + xy,
\end{aligned}
\tag{8.18}
$$

where a, b, and r are positive constants.

We could proceed with the variables x, y, and z, but to make things a bit simpler, set $u_1 = x$, $u_2 = y$, and $u_3 = z$. With these substitutions, system (8.18) takes the following vector form.

$$
\begin{bmatrix} u_1 \\ u_2 \\ u_3 \end{bmatrix}' = \begin{bmatrix} -au_1 + au_2 \\ ru_1 - u_2 - u_1 u_3 \\ -bu_3 + u_1 u_2 \end{bmatrix}
\tag{8.19}
$$

If we let $\mathbf{F}(t, [u_1, u_2, u_3]^T) = [-au_1 + au_2, ru_1 - u_2 - u_1 u_3, -bu_3 + u_1 u_2]^T$ and $\mathbf{u} = [u_1, u_2, u_3]^T$, then $\mathbf{u}' = [u_1', u_2', u_3']^T$ and system (8.19) takes the form $\mathbf{u}' = \mathbf{F}(t, \mathbf{u})$, prompting one to write the ODE file as follows.

```
function uprime = F(t,u)
uprime = zeros(3,1);
uprime(1) = -a*u(1) + a*u(2);
uprime(2) = r*u(1) - u(2) - u(1)*u(3);
uprime(3) = -b*u(3) + u(1)*u(2);
```

However, there is a big problem with this ODE file. MATLAB executes its functions, including those defined in function M-files, in separate workspaces disjoint from the command window workspace and

114

from the workspaces of other functions. The variables defined in one workspace are not available in another unless they are passed to it explicitly. The impact of this is that we will have to pass the parameters a, b, and r from the command window to the function ODE file.

One way to do this is to rewrite the ODE file as[12]

```
function uprime = lor1(t,u,a,b,r)
uprime = zeros(3,1);
uprime(1) = -a*u(1) + a*u(2);
uprime(2) = r*u(1) - u(2) - u(1)*u(3);
uprime(3) = -b*u(3) + u(1)*u(2);
```

We've decided to name our function `lor1` instead of F, so we must save the file as `lor1.m`[13].

We will use $a = 10$, $b = 8/3$, $r = 28$, and the initial condition

$$\mathbf{u}(0) = \begin{bmatrix} u_1(0) \\ u_2(0) \\ u_3(0) \end{bmatrix} = \begin{bmatrix} 1 \\ 2 \\ 3 \end{bmatrix}.$$

The commands[14]

```
>> [t,u] = ode45(@lor1,[0,7],[1;2;3],[],10,8/3,28);
>> plot(t,u)
>> title('A solution to the Lorenz system')
>> xlabel('t'), ylabel('x, y, and z')
>> legend('x','y','z'), grid
```

should produce an image similar to that in Figure 8.9. [15] Notice the use of [] as an entry in the `ode45` command. This is just a place holder. This place in the order of entries is reserved for options. We will explain these later.

Global Variables. There is another way of passing variables to function M-files that you might find more to your liking. This is the use of *global variables*. The only hard thing about the use of global variables is to remember that they must be declared in the both the command window and in the function M-file. Here is an example.

---

[12] In version 5 of MATLAB the first line of this file must read `function uprime = lor1(t,u,flag,a,b,r)`. Notice the input `flag`.

[13] It would be natural to name this file `lorenz.m`, but there is already an M-file in the MATLAB directory tree with that name. If you enter `lorenz` at the MATLAB prompt you will see a solution to the Lorenz system (8.18) displayed in a very attractive manner.

[14] In version 5 of MATLAB you must invoke the solver in the first line using the syntax `[t,u]=ode45('lor1', [0,7],[1;2;3],[],10,8/3,28);`.

[15] The linestyles have been changed in Figure 8.9 to compensate for the unavailability of color.

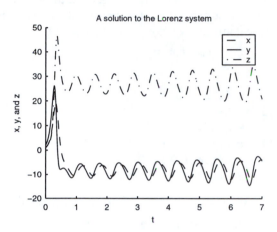

Figure 8.9. A solution to the Lorenz system.

First change the function ODE file to

```
function uprime = lor2(t,u)
global A B R
uprime = zeros(3,1);
uprime(1) = -A*u(1) + A*u(2);
uprime(2) = R*u(1) - u(2) - u(1)*u(3);
uprime(3) = -B*u(3) + u(1)*u(2);
```

Save this second version as `lor2.m`. The second line declares the parameters to be global variables. Notice that we are using uppercase names for global variables. This is good practice, but is not required.

Next, we declare the global variables in the command window and initialize them

```
>> global A B R
>> A = 10; B = 8/3; R = 28;
```

The global variables A, B, and R are now available in both the command window and in the workspace of the ODE file. You can find the numerical solution of the Lorenz system with the command[16]

```
>> [t,u] = ode45(@lor2,[0,7],[1;2;3]);
```

Eliminating Transient Behavior.

In many mechanical and electrical systems we experience *transient behavior*. This involves behavior that is present when the system starts, but dies out quickly in time, leaving only a fairly regular *steady-state* behavior. In Figure 8.9 there appears to be transient behavior until about $t = 1$, and then things seem

---

[16] In version 5 of MATLAB use `[t,u] = ode45('lor2',[0,7],[1;2;3])`.

116

to settle down. To examine the situation more closely, let's compute the solution over a longer period, say $0 \leq t \leq 100$, and then plot the part corresponding to $t > 10$ in three dimensions. We would also like to allow for randomly chosen initial values in order to see if the steady-state behavior is somehow independent of the initial conditions.

This is easily accomplished. The hardest part is choosing the initial conditions, and we can use the MATLAB command `rand` for this. If you use `help rand` you will see that `rand` produces random numbers in the range [0, 1]. We want each component of our initial conditions to be randomly chosen from the interval [−50, 50]. It is necessary to scale the output from `rand` to this interval. The command `u0 = 100*(rand(3,1) - 0.5)` will do the job.

Since we will want to repeat our experiment several times, we will put everything, including the ODE file, into a master function M-file named `loren.m`. We will use global variables initialized in both the master file and the subfunction. Including formatting commands, our file is

```
function loren

   global A B R
   A = 10; B = 8/3; R = 28;
   u0 = 100*(rand(3,1) - 0.5);
   [t,u] = ode45(@lor2,[0,100],u0);
   N = find(t>10);
   v = u(N,:);
   plot3(v(:,1),v(:,2),v(:,3))
   title('The Lorenz attractor')
   xlabel('x'), ylabel('y'), zlabel('z')
   grid, shg

function uprime = lor2(t,u)
   global A B R
   uprime = zeros(3,1);
   uprime(1) = -A*u(1) + A*u(2);
   uprime(2) = R*u(1) - u(2) - u(1)*u(3);
   uprime(3) = -B*u(3) + u(1)*u(2);
```

Further explanation is necessary. First, the command `N = find(t>10);` produces a list of the indices of those elements of t which satisfy $t > 10$. Then `v = u(N,:);` produces a matrix containing the rows of u with indices in N. The result is that v consists of only those solution points corresponding to $t > 10$, and the resulting plot represents the long time behaviour of the solution after the transient effects are gone.

Now, every time we execute `loren` we get a result very like Figure 8.10. The object in this figure is an approximation of the *Lorenz attractor*. The Lorenz system, with this particular choice of parameters, has the property that any solution, no matter what its initial point, is attracted to this rather complicated, butterfly-shaped, set. We will return to this in the exercises. There we will also examine the Lorenz system for different values of the parameters.

Click on the rotate icon in the toolbar, and click and drag the mouse in the axes in Figure 8.10 to

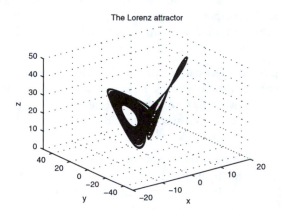

Figure 8.10. The Lorenz attractor.

rotate the axes to a different view of the Lorenz attractor. If the animation is unbearably slow, go back and repeat the commands for the construction of the Lorenz attractor, only use a smaller time interval, say [0, 20].

For an animated plot, add the line `figure(1)` after the first `function` statement, and the line

```
comet3(v(:,1),v(:,2),v(:,3))
```

immediately before the `plot` command.

Improving Accuracy

It has been mentioned before that `ode45` chooses its own step size in order to achieve a predetermined level of accuracy. Although the default level is sufficient for many problems, sometimes it will be necessary to improve that "predetermined level of accuracy."

The accuracy is controlled by the user by defining two optional inputs to `ode45`, and the other solvers. These are the *relative tolerance*, RelTol, and the *absolute tolerance*, AbsTol. If $\mathbf{y}^k$ is the computed solution at step k, then each component of the solution is required to satisfy its own error restriction. This means that we consider an estimated error vector, which has a component for every component of $\mathbf{y}^k$, and it is required that for each j, the jth component of the error vector satisfy

$$|\text{estimated error}_j^k| \leq \text{tolerance}_j, \qquad (8.20)$$

where

$$\text{tolerance}_j = \max(|y_j^k| \times \text{RelTol}, \text{AbsTol}_j). \qquad (8.21)$$

A little algebra discloses that

$$\text{tolerance}_j = \begin{cases} \text{AbsTol}_j, & \text{if } |y_j^k| \leq \text{AbsTol}_j/\text{RelTol}, \\ |y_j^k| \times \text{RelTol}, & \text{otherwise}. \end{cases} \qquad (8.22)$$

118

Thus the relative tolerance controls the tolerance unless $|y_j^k|$ is relatively small, where "relatively small" means $|y_j^k| \leq \text{AbsTol}_j/\text{RelTol}$.

Notice that the relative tolerance is a number, but the absolute tolerance is a vector quantity, with a component for each equation in the system being solved. This allows the user to set the tolerance differently for different components of the solution. The default values for RelTol is 10^{-3}, and AbsTol is a vector, each component of which is 10^{-6}. It is also possible to use a single number for the absolute tolerance, if the same tolerance will work for each component.

Choosing the Tolerances.   The philosophy behind the use of the two tolerances is that the relative tolerance should bound the estimated error by a certain fraction of the size of the computed quantity y_j^k for most values of that quantity. The absolute tolerance should come into play only for values of y_j^k which are small in comparison to the range of values that y_j^k assumes. This is to prevent the routine from trying too hard if the computed quantity turns out to be very small at some steps. For example, with the defaults of RelTol $= 10^{-3}$ and AbsTol $= 10^{-6}$, we have tolerance$_j = \text{AbsTol}_j$ only when $|y_j^k| \leq 10^{-3}$.

Let's talk about choosing the relative tolerance first. This is the easier case since we are simply talking about the number of significant digits we want in the answer. For example, the default relative tolerance is 10^{-3}, or 0.1%. In an ideal world this will ensure that the first three digits of the computed answer will be correct. If we want five significant digits we should choose 10^{-5}, or 0.001%.

It would be nice if things were as simple as these considerations make them seem. Unfortunately the inequalities in (8.20), (8.21), and (8.22) at best control the errors made at an individual step. These errors propagate, and errors made at an early step can propagate to something much larger. In fact, the errors can propagate exponentially. The upshot is that over long time periods computational solutions can develop serious inaccuracies. It is usually a good idea is to introduce a safety factor into your choice of tolerances. For example if you really want three significant digits, it might be good to choose the relative tolerance smaller than 10^{-3}, say 10^{-5}. Even then it is imperative that you check your answer with what you expect and reduce the tolerance even further if it seems to be called for. (See Example 5.)

To choose the absolute tolerance vector, remember that the philosophy is that the tolerance should be equal to the absolute tolerance only when the computed value y_j^k is small relative to the range of values of that component. Suppose that we expect that $|y_j^k| \leq M_j$ at all computed points. Then we might demand that AbsTol dominate the tolerance only when $|y_j^k| \leq 10^{-3} \times M_j$. From (8.22) we see that we want $\text{AbsTol}_j/\text{RelTol} = 10^{-3} \times M_j$, so we set

$$\text{AbsTol}_j = 10^{-3} \times M_j \times \text{RelTol}. \qquad (8.23)$$

Admittedly, it is usually impossible to know M_j before the solution is computed. However, one gains information about this as one studies the solution. Of course the factor of 10^{-3} can be replaced by a smaller or larger number if the situation at hand merits it.

Let's use these considerations to decide what tolerances to use for the Lorenz equations. Since we are only going to plot the solution, two place accuracy along the entire solution should be sufficient. Thus, the default relative tolerance of 10^{-3} would seem to be sufficient, but if we use a safety factor of 10^{-2} we get a relative tolerance of 10^{-5}. If we were to compute over longer periods of time, we might want to choose a smaller value, but without examining the computed results a relative tolerance smaller than 10^{-5} does not seem justified.

To choose the absolute tolerance vector, we notice from Figures 8.9 and 8.10 that the normal ranges of the components are $|x| \leq 20$, $|y| \leq 30$, and $|z| \leq 40$. Hence according to (8.23), we should choose

$$\text{AbsTol}_1 \leq 10^{-3} \times 20 \times 10^{-3} = 2 \times 10^{-5},$$
$$\text{AbsTol}_2 \leq 10^{-3} \times 30 \times 10^{-3} = 3 \times 10^{-5},$$
$$\text{AbsTol}_3 \leq 10^{-3} \times 40 \times 10^{-3} = 4 \times 10^{-5}.$$

We see that the default absolute tolerance of 10^{-6} for each seems to be adequate.

In our approach to choosing tolerances everything depends on choosing the relative tolerance. From our earlier discussion this seems simple enough. However, there are situations in which more accuracy may be required than appears on the surface. Suppose, for example, that it is the difference of two components of the solution which is really important. To be more specific, suppose we compute a solution $\mathbf{y}$, and what we are really interested in is $y_1 - y_2$, the difference of the first two components of $\mathbf{y}$. Suppose in addition that y_1 and y_2 are very close to each other. Suppose in fact that we know they will always agree in the first 6 digits. If we set the relative tolerance to 10^{-7}, then in an ideal world we will get 7 significant digits for y_1 and y_2. For example we might get $y_1 = 1234569*****$ and $y_2 = 1234563*****$, where the asterisks indicate digits that are not known accurately. Then $y_1 - y_2 = 6*****$, so we only know 1 significant digit in the quantity we are really interested in. Consequently to get more significant digits in $y_1 - y_2$ we have to choose a smaller relative tolerance. To get 3 significant digits we need 10^{-9}, or with a suitable safety factor, 10^{-11}.

Using the Tolerances in the Solvers. Having chosen our tolerances, how do we use them in ode45 and the other MATLAB solvers? These solvers have many options, many more than we will describe in this book. All of the options are input into the solver using an options vector. Suppose, for example, that we wanted to solve the Lorenz system with relative tolerance 10^{-5} and absolute tolerance vector $[10^{-8}, 10^{-7}, 10^{-7}]$. We build the options vector with the command odeset, and then enter it as a parameter to ode45. If we use the first version of the Lorenz ODE file, lor1.m, then the needed commands are

```
>> options  =  odeset('RelTol',1e-5,'AbsTol',[1e-8 1e-7 1e-7]);
>> [t,u] = ode45(@lor1,[0,100],[1;2;3],options,10,8/3,28);
```

If the run takes inordinately long, try a shorter time span, say [0,20] instead of [0,100]. Notice that the syntax used with ode45 is exactly the same as was used earlier for the Lorenz system, except that we have inserted the parameter options between the initial conditions and the parameters sent to the routine.

If you wish to use global variables, then use the second version of the Lorenz ODE file, lor2, with

```
>> options  =  odeset('RelTol',1e-5,'AbsTol',[1e-8 1e-7 1e-7]);
>> [t,u] = ode45(@lor2,[0,100],[1;2;3],options);
```

Again, this assumes that you have declared the global variables and initialized them.

An experimental approach to choosing the relative tolerance. Perhaps the best way to choose the relative tolerance is illustrated by the next example.

120

Example 5. *The initial value problem*[17]

$$y'' + 0.1y' + \sin y = \cos t \quad \text{with} \quad y(0) = 0 \quad \text{and} \quad y'(0) = 2$$

models a forced damped pendulum. Use ode45 *to solve the problem over the interval* [0, 150].

The homogeneous problem has stable equilibria where y is an integer multiple of 2π and $y' = 0$. The relatively large initial velocity, and the driving force can lead to erratic behavior. Small computing errors can lead to incorrect solutions, as we will see. The problem is to choose an appropriate relative tolerance for ode45.

This time we will take an experimental approach to the problem. We will solve the problem with a tolerance reltol, and then solve it again with tolerance reltol/10. We will compare the two solutions graphically, and only when they are sufficiently close will we be satisfied. The function M-file borrelli.m will do this for us.

```
function borrelli(reltol)
  close all
  tspan = [0,150];
  velocity = 2; position = 0;
  color = 'r';
  hold on
  for kk = 1:2
    opts = odeset('RelTol',reltol);
    [t,y]=ode45(@dfile,tspan,[position;velocity],opts);
    plot(t,y(:,1),'-','color',color)
    reltol = reltol/10;
    color = 'k';
  end
  hold off
  xlabel('t'), ylabel('y')
  title('y'''' + 0.1y'' + sin(y) = cos(t), y(0)=0, y''(0)=2')
  set(gca,'YTick',-2*pi:2*pi:6*pi,...
          'YTickLabel',{'-2pi','0','2pi','4pi','6pi'},...
          'YGrid','on')

function yprime=dfile(t,y)
  yprime=zeros(2,1);
  yprime(1) = y(2);
  yprime(2) = -sin(y(1)) - 0.1*y(2) + cos(t);
```

The file needs some explanation. The command close all closes all figures, and allows the routine to start with a clean slate each time it is used. Use this command with care, since it may close a figure

---

[17] For more about this example, see Robert Borrelli, Courtney Coleman *Computers, Lies, and the Fishing Season*, College Math Journal Volume 25, Number 5 (1994) and John Hubbard, *The Forced Damped Pendulum: Chaos, Complication, and Control* The American Mathematical Monthly (October 1999)

that you want to use later. The `for` loop computes and plots the solution twice. The first time it uses the input `reltol` and plots the curve in red. The second time through it uses `reltol/10` and plots the curve in black. We will be able to easily compare the graphs and we will consider the input `reltol` to be successful if very little of the red curve shows through the black curve. Finally, the `set` command sets the ticks on the y-axis to be multiples of π.

We first execute `borrelli(1e-3)`. Remember that 10^{-3} is the default relative tolerance for ode45. The result is shown in Figure 8.11. The solutions agree pretty well up to about $t = 25$ and then diverge dramatically. Clearly the default relative tolerance is not sufficiently small. When we execute `borrelli(1e-4)` we get Figure 8.12. This time the curves overlap, so using a relative tolerance of 10^{-4} will do an adequate job, although 10^{-5} would be a more conservative choice.

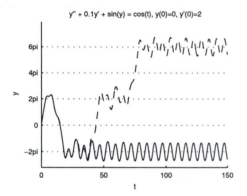

Figure 8.11. The solutions with relative tolerances 10^{-3} (solid) and 10^{-4} (dashed).

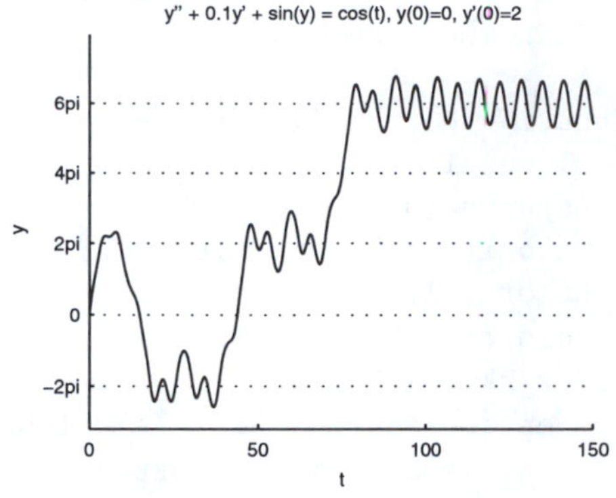

Figure 8.12. The solutions with relative tolerances 10^{-4} (solid) and 10^{-5} (dashed) overlap.

Example 5 illustrates a procedure that should be followed whenever you compute the solution to a system of differential equations over a relatively long interval. Compare the results with a chosen relative tolerance and with that divided by 10. Reduce the relative tolerance until the two solutions agree. Only then can you have confidence in your solution.

Behavior Near Discontinuities

Any numerical method will run into difficulties where the equation and/or the solution to the equation has a point of discontinuity, and ode45 is no exception. There are at least three possible outcomes when ode45 meets a discontinuity.

1. ode45 can integrate right through a discontinuity in the equation or the solution, not even realizing it is there. In this case the accuracy of the result is highly doubtful, especially in that range beyond the discontinuity. This phenomenon will happen, for example, with the initial value problem $x' = (x + 3t^2)/t$, with initial value $x(-1) = 1$ on the interval $[-1, 1]$. Another example is $x' = t/(x + 1)$, with $x(-2) = 0$ on the interval $[-2, -3/2]$.
2. ode45 can find the discontinuity and report it with a warning. This is usually the sign that there is

122

a discontinuity in the solution. An example of this is $x' = x^2$, with $x(0) = 1$ on the interval $[0, 2]$. The solution is $x(t) = 1/(1-t)$, which has a singularity at $t = 1$. When asked to solve this initial value problem, ode45 responds with

```
Warning: Failure at t=9.999694e-001.  Unable to meet integration
  tolerances without reducing the step size below the smallest
  value allowed (3.552605e-015) at time t.
```

One very nice thing about ode45 is that if this happens, the output up to the point where the routine stops is made available. For example, if you execute `[t,w]  = ode45('gizmo',[0,3.4],[0; 1; 2]);`, and the computation stops because a step size is called for which is smaller than the minimum allowed, then the variables t and w will contain the results of the computation up to that point.

3. ode45 can choose smaller and smaller step sizes upon the approach to a discontinuity and go on calculating for a very long time — hours in some cases. For cases like this, it is important for the user to know how to stop a calculation in MATLAB. On most computers the combination of the control key and C depressed simultaneously will do the trick. An example is $x' = t/(x + 1)$, with $x(-10) = 2$ on the interval $[-10, -5]$. Many stiff equations and systems will also require an inordinate solution time using ode45.

Stiff Equations

Solutions to differential equations often have components which are varying at different rates. If these rates differ by a couple of orders of magnitude, the equations are called stiff. Stiff equations will not arise very often in this manual, but we will say a word about them in case you should have to deal with one. We will also put some examples in the exercises.

The equation $x' = e^t \cos x$ from Example 2 is stiff over relatively large time intervals, simply because the factor e^t gets very large. To see what this does to ode45, execute

```
>> f = inline('exp(t)*cos(x)','t','x');
>> ode45(f,[0,10],0)
```

Since we have not required any output from ode45, it will plot the solution as it computes it. You will notice that the speed of the plot slows up dramatically as t increases. When you are tired of waiting for the solution to end, click on the **Stop** button on the lower left of the figure window. This very slow computation is the typical response of ode45 to a stiff equation or system. The fast reaction rate, measured in this case by the factor e^t, requires ode45 to take extremely small steps, and therefore a long time to complete the solution.

The MATLAB suite of ode solvers contains four routines, ode15s, ode23s, ode23t, and ode23tb, which are designed to solve stiff equations. ode15s is the first of these to try. To see what it does in this case, execute

```
>> ode15s(f,[0,10],0)
```

Since it is designed specifically for stiff systems, it completes its work in seconds. Notice that the calling syntax for ode15s is identical to that of ode45. This is a feature of all of the solvers in the ODE suite.

Another example of a stiff system is the van der Pol system

$$x' = y,$$
$$y' = -x + \mu(1 - x^2)y,$$

when the parameter μ is very large, say $\mu = 1000$.

But how do we tell if a system is stiff? It is often obvious from the physical situation being modeled that there are components of the solution which vary at rates that are significantly different, and therefore the system is stiff. Sometimes the presence of a large parameter in the equations is a tip off that the system is stiff. However, there is no general rule that allows us to recognize stiff systems. A good operational rule is to try to solve using `ode45`. If that fails, or is very slow, try `ode15s`.

Other Possibilities

The MATLAB ODE Suite consists of a very powerful group of solvers. We have only touched the surface of the options available to control them, and the variations of the ways in which they can be used. To mention just a few, it is possible to modify the output of the solver so that the solution is automatically plotted in the way the user desires. It is also possible to get the solvers to detect events. For example the 3-body problem of the sun, earth, and moon can be modeled by a system of 12 ODEs, and the solver can be set up to automatically detect the times of eclipses of the sun and moon.

A very nice description of the ODE Suite appears in *Using MATLAB*, which is part of the MATLAB documentation. You should be able to find it by entering `helpdesk` at the MATLAB prompt. After the MATLAB Help Desk opens, click on Using MATLAB, then on Mathematics, and finally on Differential Equations.

A more technical paper dealing the the ODE Suite is also available from the Online Manuals list. It is written by Larry Shampine and Mark Reichelt, who built the Suite.

Exercises

In Exercises 1 – 3 we will deal with the differential equation $y' = -2y + 2\cos t \sin 2t$. In preparation create the function M-file

```
function yprime = steady(t,y)
yprime = -2*y+2*cos(t).*sin(2*t);
```

and save the file as `steady.m`.

1. MATLAB's solvers are able to handle multiple initial conditionsfor single first order equations. To do so we fake `ode45` into assuming that we are solving a system consisting of the same equation repeated for each initial condition. For this purpose it is only necessary to make the function ODE file array smart, as we have for `steady.m`.

   a) To solve the equation $y' = -2y + 2\cos t \sin 2t$, for initial conditions $y(0) = -5$, $y(0) = -4,\ldots$, $y(0) = 5$, enter the code `[t,y] = ode45(@steady,[0,30],-5:5);`.

   b) Graph the solutions with `plot(t,y)` and note that each solution approaches a periodic steady-state solution.

   c) Repeat parts a) and b) with time spans $[0, 10]$ and $[0, 2\pi]$ to get a closer look at the convergence to the steady-state solution.

2. Another nice feature of MATLAB's solvers is the ability to output solutions at specified times. This is useful when you want to compare two functions at specified time points or place the output from multiple calls to ode45 in a matrix, where it is required that the rows have equal length. Enter [t,y] = ode45(@steady,0:.25:3,1);, then enter [t,y]. This should clarify what is going on.

3. MATLAB's solvers can automatically plot solutions *as they are computed.* All you need to do is leave off the output variables when calling the solver. Try ode45('steady',[0,30],-5:5).

In Exercises 4 – 7, plot the solutions to all of the initial value problems on one figure. Choose a solution interval starting at $t = 0$ that well illustrates any steady-state phenomena that you see. Exercise 1 might be helpful.

4. $y' + 4y = 2\cos t + \sin 4t$, with $y(0) = -5, -4, \ldots, 5$.

5. $y' + 4y = t^2$, with $y(0) = -9, -8, -7, -1, 1, 7, 8, 9$.

6. $x' = \cos t - x^3$, with $x(0) = -3, -2, \ldots, 3$.

7. $x' + x + x^3 = \cos^2 t$, with $x(0) = -3, -2, \ldots, 3$.

8. Separate variables and find an explicit solution of $x' = t/(1 + x)$, $x(0) = 1$.

   a) Use ode45 to find the numerical solution of the initial value problem on the interval $[0, 3]$. Store the solution in the variables t and x_ode45.

   b) Use the explicit solution to find exact solution values at each point in the vector t obtained in part a). Store the results in the variable x_exact. Compare the solutions graphically with plot(t, x_exact, t, x_ode45,'r.'). Obtain a printout with a legend.

   c) For a finer graphical image of the error plot x_exact - x_ode45 versus t in a new figure. Pay special attention to the dimensions on the y-axis.

In Exercises 9 – 12, write a function ODE file for the system. Solve the initial value problem using ode45. Provide plots of x_1 versus t, x_2 versus t, and x_2 versus x_1 on the time interval provided. See Exercises 1 – 6 in Chapter 7 for a nice method to arrange these plots.

9. $x_1' = x_2$ and $x_2' = (1 - x_1{}^2)x_2 - x_1$, with $x_1(0) = 0$ and $x_2(0) = 4$ on $[0, 10]$.

10. $x_1' = x_2$ and $x_2' = -25x_1 + 2\sin 4t$, with $x_1(0) = 0$ and $x_2(0) = 2$ on $[0, 2\pi]$.

11. $x_1' = (x_2 + x_1/5)(1 - x_1^2)$ and $x_2' = -x_1(1 - x_2^2)$, with $x_1(0) = 0.8$ and $x_2(0) = 0$ on $[0, 30]$.

12. $x_1' = x_2$ and $x_2' = -x_1 + x_2(1 - 3x_1^2 - 3x_2^2)$, with $x_1(0) = 0.2$ and $x_2(0) = 0.2$ on $[0, 20]$.

13. The built-in plotting routines of MATLAB's solvers were illustrated in Exercise 3. They work equally well with systems. For example, create

```
function yprime = heart(t,y)
yprime = zeros(2,1);
yprime(1) = y(2);
yprime(2) = -16*y(1)+4*sin(2*t);
```

and save as heart.m.

   a) Enter ode45(@heart,[0,2*pi],[0;2]) and note that MATLAB dynamically plots both y_1 and y_2 versus t.

   b) To change the output routine, enter options = odeset('OutputFcn','odephas2'), followed by ode45(@heart,[0,2*pi],[0;2],options). Note that MATLAB dynamically plots y_2 versus y_1 in the phase plane.

14. Use the technique of Example 4 to change the initial value problem $yy'' - (y')^2 - y^2 = 0$, $y(0) = 1$, $y'(0) = -1$ to a system with initial conditions. Use ode45 to create a plot of y versus t.

15. An unforced spring-mass system with no damping can be modeled by the equation $y'' + ky = 0$, where k is the spring constant. However, suppose that the spring loses elasticity with time. A possible model for an "aging" spring-mass system is $y'' + 2e^{-0.12t}y = 0$. Suppose that the spring is stretched two units from equilibrium and released from rest. Use the technique of Example 4 to plot a solution of y versus t on the time interval $[0, 100]$.

16. The equation $x'' = ax' - b(x')^3 - kx$ was devised by Lord Rayleigh to model the motion of the reed in a clarinet. With $a = 5$, $b = 4$, and $k = 5$, solve this equation numerically with initial conditions $x(0) = A$, and $x'(0) = 0$ over the interval $[0, 10]$ for the three choices $A = 0.5$, 1, and 2. Use MATLAB's hold command to prepare both time plots and phase plane plots containing the solutions to all three initial value problems superimposed. Describe the relationship that you see between the three solutions.

17. Consider the forced, damped oscillator $y'' + 0.05y' + 36y = \sin 6.3t$, with initial conditions $y(0) = 5$, $y'(0) = 0$. Use the technique of Example 4 to transform the equation to system of first order ODEs, write and save the odefile as `osc.m`, and solve the system with the command `[t,x] = ode45(@osc,[0,100],[5;0]);`.

    a) The command `plot3(x(:,1),x(:,2),t)` provides a three dimensional plot with time on the vertical axis.

    b) The command `close all, comet3(x(:,1),x(:,2),t)` provides an animated view of the result in part a).

18. Consider the forced, undamped oscillator modeled by the equation $y'' + \omega_0^2 y = 2\cos\omega t$. The parameter ω_0 is the *natural frequency* of the unforced system $y'' + \omega_0^2 y = 0$. Write an ODE function M-file for this equation with natural frequency $\omega_0 = 2$, and the driving frequency ω passed as a parameter.

    a) If the driving frequency nearly matches the natural frequency, then a phenomenon called *beats* occurs. As an example, solve the equation using ode45 with $y(0) = 0$ and $y'(0) = 0$ over the interval $[0, 60\pi]$ with $\omega = 1.9$.

    b) If the driving frequency matches the natural frequency, then a phenomenon called *resonance* occurs. As an example, solve the equation using ode45 with $y(0) = 0$ and $y'(0) = 0$ over the interval $[0, 60\pi]$ with $\omega = 2$.

    c) Use the `plot3` and `comet3` commands, as described in Exercise 17, to create three dimensional plots of the output in parts a) and b).

19. **Harmonic motion.** An unforced, damped oscillator is modeled by the equation

$$my'' + cy' + ky = 0,$$

where m is the mass, c is the damping constant, and k is the spring constant. Write the equation as a system of first order ODEs and create a function ODE file that allows the passing of parameters m, c, and k. In each of the following cases compute the solution with initial conditions $y(0) = 1$, and $y'(0) = 0$ over the interval $[0, 20]$. Prepare both a plot of y versus t and a phase plane plot of y' versus y.

    a) (No damping) $m = 1$, $c = 0$, and $k = 16$.

    b) (Under damping) $m = 1$, $c = 2$, and $k = 16$.

    c) (Critical damping) $m = 1$, $c = 8$, and $k = 16$.

    d) (Over damping) $m = 1$, $c = 10$, and $k = 16$.

20. The system

$$\varepsilon \frac{dx}{dt} = x(1-x) - \frac{(x-q)}{(q+x)} fz,$$

$$\frac{dz}{dt} = x - z,$$

models a chemical reaction called an *oregonator*. Suppose that $\varepsilon = 10^{-2}$ and $q = 9 \times 10^{-5}$. Put the system into normal form and write an ODE function M-file for the system that passes f as a parameter. The idea is to vary the parameter f and note its affect on the solution of the oregonator model. We will use the initial conditions $x(0) = 0.2$ and $y(0) = 0.2$ and the solution interval $[0, 50]$.

    a) Use ode45 to solve the system with $f = 1/4$. This should provide no difficulties.

    b) Use ode45 to solve the system with $f = 1$. This should set off all kinds of warning messages.

    c) Try to improve the accuracy with `options = odeset('RelTol',1e-6)` and using options as the options parameter in ode45. You should find that this slows computation to a crawl as the system is very stiff.

d) The secret is to use `ode15s` instead of `ode45`. Then you can note the oscillations in the reaction.

21. Consider the three-dimensional oregonator model

$$\frac{dx}{dt} = \frac{qy - xy + x(1 - x)}{\varepsilon},$$

$$\frac{dy}{dt} = \frac{-qy - xy + fz}{\varepsilon'},$$

$$\frac{dz}{dt} = x - z.$$

Set $\varepsilon = 10^{-2}$, $\varepsilon' = 2.5 \times 10^{-5}$, and $q = 9 \times 10^{-5}$. Set up an ODE file that will allow the passing of the parameter f. Slowly increase the parameter f from a low of 0 to a high of 1 until you clearly see the system oscillating on the interval $[0, 50]$ with initial conditions $x(0) = 0.2$, $y(0) = 0.2$, and $z(0) = 0.2$. You might want to read the comments in Exercise 20. Also, you'll want to experiment with `plot`, `zoom`, and `semilogy` to best determine how to highlight the oscillation in all three components.

22. The system

$$m_1 x'' = -k_1 x + k_2(y - x),$$

$$m_2 y'' = -k_2(y - x),$$

models a *coupled* oscillator. Imagine a spring (with spring constant k_1) attached to a hook in the ceiling. Mass m_1 is attached to the spring, and a second spring (with spring constant k_2) is attached to the bottom of mass m_1. If a second mass, m_2, is attached to the second spring, you have a coupled oscillator, where x and y represent the displacements of masses m_1 and m_2 from their respective equilibrium positions.

If you set $x_1 = x$, $x_2 = x'$, $x_3 = y$, and $x_4 = y'$, you can show that

$$x_1' = x_2,$$

$$x_2' = -\frac{k_1}{m_1} x_1 + \frac{k_2}{m_1}(x_3 - x_1),$$

$$x_3' = x_4,$$

$$x_4' = -\frac{k_2}{m_2}(x_3 - x_1).$$

Assume $k_1 = k_2 = 2$ and $m_1 = m_2 = 1$. Create an ODE file and name it `couple.m`. Suppose that the first mass is displaced upward two units, the second downward two units, and both masses are released from rest. Plot the position of each mass versus time.

23. In the 1920's, the Italian mathematician Umberto Volterra proposed the following mathematical model of a predator-prey situation to explain why, during the first World War, a larger percentage of the catch of Italian fishermen consisted of sharks and other fish eating fish than was true both before and after the war. Let $x(t)$ denote the population of the prey, and let $y(t)$ denote the population of the predators.

In the absence of the predators, the prey population would have a birth rate greater than its death rate, and consequently would grow according to the exponential model of population growth, i.e. the growth rate of the population would be proportional to the population itself. The presence of the predator population has the effect of reducing the growth rate, and this reduction depends on the number of encounters between individuals of the two species. Since it is reasonable to assume that the number of such encounters is proportional to the number of individuals of each population, the reduction in the growth rate is also proportional to the product of the two populations, i.e., there are constants a and b such that

$$x' = ax - bxy. \tag{8.24}$$

Since the predator population depends on the prey population for its food supply it is natural to assume that in the absence of the prey population, the predator population would actually decrease, i.e. the growth rate would be negative. Furthermore the (negative) growth rate is proportional to the population. The presence of the prey population would provide a source of food, so it would increase the growth rate of the predator species. By the same reasoning used for the prey species, this increase would be proportional to the product of the two populations. Thus, there are constants c and d such that

$$y' = -cy + dxy. \tag{8.25}$$

a) A typical example would be with the constants given by $a = 0.4$, $b = 0.01$, $c = 0.3$, and $d = 0.005$. Start with initial conditions $x_1(0) = 50$ and $x_2(0) = 30$, and compute the solution to (8.22) and (8.23) over the interval [0, 100]. Prepare both a time plot and a phase plane plot.

After Volterra had obtained his model of the predator-prey populations, he improved it to include the effect of "fishing," or more generally of a removal of individuals of the two populations which does not discriminate between the two species. The effect would be a reduction in the growth rate for each of the populations by an amount which is proportional to the individual populations. Furthermore, if the removal is truly indiscriminate, the proportionality constant will be the same in each case. Thus, the model in equations (8.22) and (8.23) must be changed to

$$\begin{aligned} x' &= ax - bxy - ex, \\ y' &= -cy + dxy - ey, \end{aligned} \tag{8.26}$$

where e is another constant.

b) To see the effect of indiscriminate reduction, compute the solutions to the system in (8.24) when $e = 0$, 0.01, 0.02, 0.03, and 0.04, and the other constants are the same as they were in part a). Plot the five solutions on the same phase plane, and label them properly.

c) Can you use the plot you constructed in part b) to explain why the fishermen caught more sharks during World War I? You can assume that because of the war they did less fishing.

Student Projects

The following problems are quite involved, with multiple parts, and thus are good candidates for individual or group student projects.

1. **The Non-linear Spring and Duffing's Equation.**

A more accurate description of the motion of a spring is given by *Duffing's equation*

$$my'' + cy' + ky + ly^3 = F(t).$$

Here m is the mass , c is the damping constant, k is the spring constant, and l is an additional constant which reflects the "strength" of the spring. Hard springs satisfy $l > 0$, and soft springs satisfy $l < 0$. As usual, $F(t)$ represents the external force.

Duffing's equation cannot be solved analytically, but we can obtain approximate solutions numerically in order to examine the effect of the additional term ly^3 on the solutions to the equation. For the following exercises, assume that $m = 1$ kg, that the spring constant $k = 16$ N/m, and that the damping constant is $c = 1$ kg/sec. The external force is assumed to be of the form $F(t) = A\cos(\omega t)$, measured in Newtons, where ω is the frequency of the driving force. The natural frequency of the spring is $\omega_0 = \sqrt{k/m} = 4$ rad/sec.

a) Let $l = 0$ and $A = 10$. Compute the solution with initial conditions $y(0) = 1$, and $y'(0) = 0$ on the interval [0, 20], with $\omega = 3.5$ rad/sec. Print out a graph of this solution. Notice that the steady-state part of the solution dominates the transient part when t is large.

b) With $l = 0$ and $A = 10$, we want to compute the amplitude of the steady-state solution. The amplitude is the maximum of the values of $|y(t)|$. Because we want the amplitude of the steady-state oscillation, we only want to allow large values of t, say $t > 15$. This will allow the transient part of the solution to decay. To compute this number in MATLAB, do the following. Suppose that Y is the vector of y-values, and T is the vector of corresponding values of t. At the MATLAB prompt type

```
max(abs(Y.*(T>15)))
```

128

Your answer will be a good approximation to the amplitude of the steady-state solution.

Why is this true? The expression (T>15) yields a vector the same size as T, and an element of this vector will be 1 if the corresponding element of T is larger than 15 and 0 otherwise. Thus Y.*(T>15) is a vector the same size as T or Y with all of the values corresponding to $t \leq 15$ set to 0, and the other values of Y left unchanged. Hence, the maximum of the absolute values of this vector is just what we want.

Set up a script M-file that will allow you to do the above process repeatedly. For example, if the derivative M-file for Duffing's equation is called duff.m, the script M-file could be

```
[t,y] = ode45(@duff'0,20,[1,0]);
y   =  y(:,1);
amplitude  =  max(abs(y.*(t>15)))
```

Now by changing the value of ω in duff.m you can quickly compute how the amplitude changes with ω. Do this for eight evenly spaced values of ω between 3 and 5. Use plot to make a graph of amplitude vs. frequency. For approximately what value of ω is the amplitude the largest?

c) Set $l = 1$ (the case of a hard spring) and $A = 10$ in Duffing's equation and repeat b). Find the value of ω, accurate to 0.2 rad/sec, for which the amplitude reaches its maximum.

d) For the hard spring in c), set $A = 40$. With $y(0) = y'(0) = 0$, and for eight (or more) evenly spaced values of ω between 5 and 7 compute the magnitude of the steady-state oscillation. Plot the graph of amplitude vs. frequency. Repeat the computation with initial conditions $y(0) = 6$, $y'(0) = 0$, and plot the two graphs of amplitude vs. frequency on the same figure. (The phenomenon you will observe is called Duffing's hysteresis.)

e) Set $l = -1$ and $A = 10$ (the case of a soft spring) in Duffing's equation and repeat c).

2. The Lorenz System.

The purpose of this exercise is to explore the complexities displayed by the Lorenz system as the parameters are varied. We will keep $a = 10$, and $b = 8/3$, and vary r. For each value of r, examine the behavior of the solution with different initial conditions, and make conjectures about the limiting behavior of the solutions as $t \to \infty$. The solutions should be plotted in a variety of ways in order to get the information. These include time plots, such as Figure 8.9, and phase space plots, such as Figure 8.10. In the case of phase space plots, use the rotate3d command to capture a view of the system that provides the best visual information. You might also use other plots, such as z versus x, etc.

Examine the Lorenz system for a couple of values of r in each of the following intervals. Describe the limiting behavior of the solutions.

a) $0 < r < 1$.

b) $1 < r < 470/19$. There are two distinct cases.

c) $470/19 < r < 130$. The case done in the text. This is a region of chaotic behavior of the solutions.

d) $150 < r < 165$. Things settle down somewhat.

e) $215 < r < 280$. Things settle down even more.

3. Motion Near the Lagrange Points.

Consider two large spherical masses of mass $M_1 > M_2$. In the absence of other forces, these bodies will move in elliptical orbits about their common center of mass (if it helps, think of these as the earth and the moon). We will consider the motion of a third body (perhaps a spacecraft), with mass which is negligible in comparison to M_1 and M_2, under the gravitational influence of the two larger bodies. It turns out that there are five equilibrium points for the motion of the small body relative to the two larger bodies. Three of these were found by Euler, and are on the line connecting the two large bodies. The other two were found by Lagrange and are called the Lagrange points. Each of these forms an equilateral triangle in the plane of motion with the positions of the two large bodies. We are interested in the motion of the spacecraft when it starts near a Lagrange point.

In order to simplify the analysis, we will make some assumptions, and choose our units carefully. First we will assume that the two large bodies move in circles, and therefore maintain a constant distance from each other. We will take the origin of our coordinate system at the center of mass, and we will choose rotating coordinates, so that the x-axis always contains the two large bodies. Next we choose the distance between the large bodies to be the unit of distance, and the sum of the the two masses to be the unit of mass. Thus $M_1 + M_2 = 1$. Finally we choose the unit of time so that a complete orbit takes 2π units; i.e., in our units, a year is 2π units. This last choice is equivalent to taking the gravitational constant equal to 1.

With all of these choices, the fundamental parameter is the relative mass of the smaller of the two bodies

$$\mu = \frac{M_2}{M_1 + M_2} = M_2.$$

Then the location of M_1 is $(-\mu, 0)$, and the position of M_2 is $(1 - \mu, 0)$. The position of the Lagrange point is $(x_0, y_0) = ((1 - 2\mu)/2, \sqrt{3}/2)$. If (x, y) is the position of the spacecraft, then the distances to M_1 and M_2 are r_1 and r_2 where

$$r_1^2 = (x + \mu)^2 + y^2,$$
$$r_2^2 = (x - 1 + \mu)^2 + y^2.$$

Finally, Newton's equations of motion in this moving frame are

$$x'' - 2y' - x = -(1 - \mu)(x + \mu)/r_1^3 - \mu(x - 1 + \mu)/r_2^3,$$
$$y'' + 2x' - y = -(1 - \mu)y/r_1^3 - \mu y/r_2^3.$$

$$(8.27)$$

a) Find a system of four first order equations which is equivalent to system (8.25).

b) The Lagrange points are stable for $0 < \mu < 1/2 - \sqrt{69}/18) \approx 0.03852$, and in particular for the earth/moon system, for which $\mu = 0.0122$. Starting with initial conditions which are less than $1/100$ unit away from the Lagrange point, compute the solution. Write $x = (1 - 2\mu)/2 + \xi$, and $y = \sqrt{3}/2 + \eta$, so that $(\xi, \eta) = (x, y) - (x_0, y_0)$ is the position of the spacecraft relative to the Lagrange point. For each solution that you compute, make a plot of η vs. ξ to see the motion relative to the Lagrange point, and a plot of y vs. x, which also includes the positions of M_1 and M_2 to get a more global view of the motion.

c) Examine the range of stability by computing and plotting orbits for $\mu = 0.037$ and $\mu = 0.04$.

d) What is your opinion? Assuming μ is in the stable range, are the Lagrange points just stable, or are they asymptotically stable?

e) Show that there are five equilibrium points for the system you found in a). Show that there are three on the x-axis (the Euler points), and two Lagrange points as described above. This is an algebraic problem of medium difficulty. It is not a computer problem unless you can figure out how to get the Symbolic Toolbox to find the answer.

f) Show that the Euler points are always unstable. This is a hard algebraic problem.

g) Show that the Lagrange points are stable for $0 < \mu < 1/2 - \sqrt{69}/18)$ (hard). Decide whether or not these points are asymptotically stable (extremely difficult).

9. Introduction to ODESOLVE

The MATLAB function `odesolve`[1] provides a way to use MATLAB's solvers with a graphical user interface. It can be applied to any system of first order equations, and therefore to any system of any order. The dimension of the system is limited only by the size of the computer screen. Since it has a graphical interface, it is relatively easy to use, but the interface does present limitations as well.

Starting odesolve

The interface of `odesolve` is very similar to those found in `dfield` and `pplane`. To begin, execute

```
>> odesolve
```

at the MATLAB prompt. This opens the **ODESOLVE Setup** window. An example can be seen in Figure 9.2. The similarities with `dfield` and `pplane` are clear. However there are differences as well. One of the most important is the edit box for the number of equations. Change this number to 20 and depress the Tab or Enter key. The setup window expands to allow the entry of 20 equations and the corresponding initial conditions.

Example 1. *Consider the two masses m_1 and m_2 connected by springs and driven by an external force, as illustrated in Figure 9.1. In the absence of the force, each of the masses has an equilibrium position. Let x and y denote the displacements of the masses from their equilibrium positions, as shown in Figure 9.1. Plot the displacements for a variety of values of the parameters and external forces.*

Analysis of the forces and Newton's second law lead us to the system of equations

$$m_1 x'' = k_2(y - x) - k_1 x,$$
$$m_2 y'' = -k_2(y - x) + f(t), \tag{9.1}$$

where k_1 and k_2 are the spring constants, and $f(t)$ is the external force.

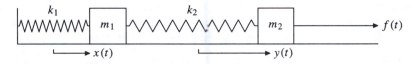

Figure 9.1. Coupled oscillators with a driving force.

---

[1] Odesolve is not part of the MATLAB distribution. Consult the appendix to Chapter 3 to learn how to obtain it.

If we introduce the velocities $u = x'$ and $v = y'$, the system in (9.1) is equivalent to the first order system

$$x' = u,$$
$$u' = (k_2(y - x) - k_1 x)/m_1,$$
$$y' = v,$$
$$v' = (-k_2(y - x) + f(t))/m_2.$$

(9.2)

This is the system we want to solve using `odesolve`. To do so, we change the number of equations to 4, and then enter the differential equations. We can speed the process by executing **Gallery→two springs**. The result is shown in Figure 9.2. Notice that the equations are entered almost like they appear in (9.2).

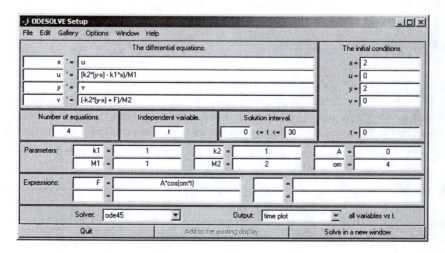

Figure 9.2. The setup window for `odesolve`.

Values for the initial conditions have been entered. The spring constants and the masses have been given values in the parameter edit boxes. The force $f(t) = A\cos(\omega t)$ has been entered as an expression, while values for the amplitude A and frequency ω have been entered in parameter boxes. In `odesolve`, the parameters must be numbers, but expressions can be any algebraic expression involving the variables and the parameters.

It is also true that the initial conditions can use the parameters, but that does not occur in this case. You will notice that we did not use TEX notation for the subscripted string constants and masses. TEX notation is allowed for variable names, but not for the names of parameters and expressions.

The independent variable has been identified as t. It has been given an initial value, and a solution interval has been designated.

Near the bottom of the setup window, we see that `ode45` is the solver that will be used. This can be changed to any of the MATLAB solvers using the popup menu. We are also told that the output will be all variables plotted against t. This can also be changed using the popup menu. For the time being let's leave these settings as they are.

To start a solution, click on the button **Solve in a new window.** After a few seconds, we get a plot of all four components of the solution to (9.2) plotted against t. Even with the legend, this figure is a little busy. We are really interested in the displacements, so let's go back to the setup window and select **time plot** from the output popup menu. Do so even though it looks like it is already selected. In the little window that appears, select only x and y, and then click **Change**. Click on **Solve in a new window** again. The result is shown in Figure 9.3.

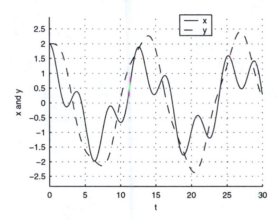

Figure 9.3. The displacements of the two springs.

Experiment with other outputs. The phase plane plot of y vs. x is very interesting, as are some of the other choices. Experiment with the use of the other solvers. Change the initial conditions and the parameters.

Example 2. *Duffing's equation,*

$$mx'' + Cx' + \omega_0^2 x + kx^3 = F(t),$$

models a forced, damped, vibrating spring where the restoring force is $-(\omega_0^2 x + kx^3)$, *a nonlinear generalization of Hooke's law. Assume that* $m = 1$, $C = 1$, $\omega_0 = 4$, *and that the external force is* $F(t) = 40\cos 6t$. *In the linear case, when* $k = 0$, *the steady state response is independent of the initial conditions. Verify this for the two sets of initial conditions* $x(0) = 0$, $x'(0) = 0$, *and* $x(0) = 6$, $x'(0) = 0$. *Next set* $k = 1$, *and examine what happens for the same initial conditions.*

If we introduce the velocity $v = x'$, then Duffing's equation is equivalent to the system

$$
\begin{aligned}
x' &= v, \\
v' &= (-Cv - \omega_0^2 x - kx^3 + F)/m.
\end{aligned}
\tag{9.3}
$$

This system should be entered into the ODESOLVE Setup window, after changing the number of equations to 2. It can also be chosen from the **Gallery** menu. After entering all of the parameters, and setting the solution interval to [0, 30], the Setup window should be similar to Figure 9.4.

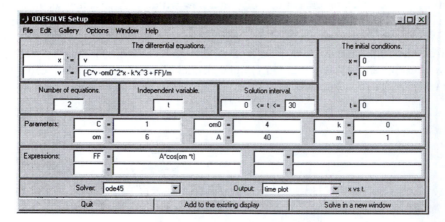

Figure 9.4. The setup window for Example 2.

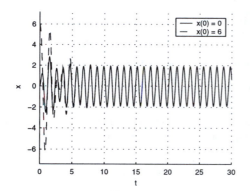

Figure 9.5. The steady-state response of a linear spring does not depend on the initial conditions.

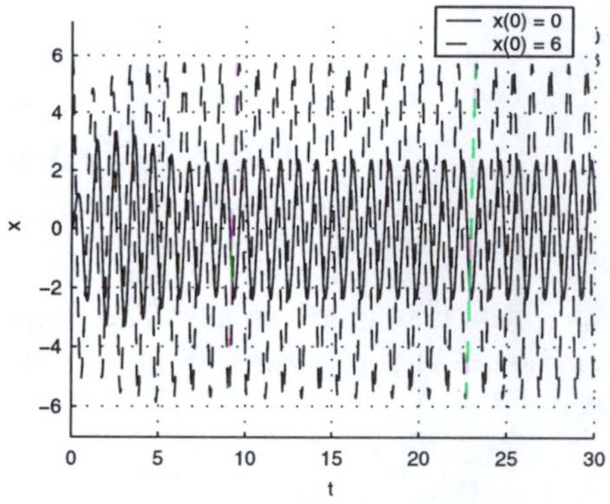

Figure 9.6. The steady-state response of a nonlinear spring does depend on the initial conditions.

When the settings are all correct for the linear spring ($k = 0$), and the initial conditions $x(0) = 0$ and $v(0) = x'(0) = 0$, click on the button **Solve in a new window.** Next go back to the Setup window, and change the initial condition to $x(0) = 6$. This time click on the button **Add to the previous display.**

The response should look similar to Figure 9.5. The legend on your plot will not be like that in Figure 9.5. To make it look the same, first use the Property Editor to change the line style of one of the graphs, and then execute

```
>> legend off
>> legend('x(0) = 0','x(0) = 6')
```

Figure 9.5 shows that the steady-state responses for these two cases are the same.

Return to the setup window and change the parameter k to 1 to introduce the nonlinearity. Then reset the initial condition to $x = 0$, and repeat the steps. The result is shown in Figure 9.6. The steady-state responses are clearly different. This is another way in which nonlinear systems behave differently from linear systems.

The **Gallery** menu contains several interesting systems. Experiment with these to learn more about how odesolve works. If you are wondering about stiff equations, choose the 1D stiff example. Try to solve using ode45. Click the Stop button when you are tired of waiting. Then solve using ode15s.

Options in ODESOLVE

If you find it difficult to see where a solution curve starts, choose the option to mark initial points. You can also choose to mark all of the points computed by the solver.

Since odesolve has a limited number of output possibilities, it will sometimes be necessary to have access to the solution data. There are two ways of doing this using odesolve. First, you can compute a solution using odesolve and export the data to the command window workspace. The data in the command window will be a structure with the name odedata*, where * is a small number chosen to not conflict with previously saved data. The structure will have a field for each variable. (See Exercises 9 and 10.)

The second method is to export an ODE function M-file, which can be used with any of MATLAB's solvers to find numerical solutions. If you choose this option, you will be asked for a filename. If you then execute help on this file, you will be given instructions on how to use a solver with this file.

There will be times when you will want to give a solver a smaller tolerance, or change some other solver setting. This is possible in odesolve by selecting **Options→Solver settings**. In this window it is possible to change settings for the solvers which we have not discussed in this chapter. Read the MATLAB documentation for the definitions.

Finally, odesolve has the capability to save and load systems and whole galleries. These options are accessed from the **File menu.** They work like the same options in dfield and pplane.

Exercises

Most of these exercises are of an advanced nature. However, any of the exercises from Chapter 8 may be done using odesolve, so look there for more elementary exercises.

1. Revisit Example 1, but do a 2D phase plot, plotting the position of the second mass versus the position of the first mass (y versus x). Extend the time period and replot. Do you think the solution is periodic?

2. Plot the solution of $x''' + xx''/50 = \cos t$, $x(0) = 1$, $x'(0) = -1$, $x''(0) = 0$ over the time interval $[0, 30]$.

3. Plot the solution of $x'''' + x'' + x = \cos t$, $x(0) = 1$, $x'(0) = -1$, $x''(0) = 0$, $x'''(0) = 0$ over the time interval $[0, 25]$. What happens to the solution if you extend the time interval?

4. A salt solution containing 2 pounds of salt per gallon of water enters a tank at a rate of 4 gal/min. Salt solution leaves the tank at a rate of 2 gal/min. Initially, the tank contains 10 gallons of pure water. Find the amount of salt in the tank when the volume of solution in the tank reaches 30 gallons. One way to do this problem is to note that the volume of solution in the tank is given by $V = 2t + 10$ and the salt content is governed by the differential equation $x' = 8 - 2x/V$, with initial condition $x(0) = 0$. Set $V = 30$ in $V = 2t + 10$ and find the time. Then solve the initial value problem and substitute this time to get the salt content. However,

`odesolve` offers another method. Set up the system

$$\frac{dx}{dt} = 8 - \frac{2x}{V}$$

$$\frac{dV}{dt} = 2.$$

Use a time plot to plot both x and V versus t for the solution with initial conditions $x(0) = 0$ and $V(0) = 10$. Examine the plot to determine when $V = 30$ and note the time when this occurs, then estimate the salt content from the plot at this time.

5. Initially, tank A contains 20 gallons of salt solution, while tanks B and C contain 40 and 50 gallons of pure water, respectively. Pure water enters tank A at a rate of 5 gal/min. Salt solution leaves tank A at a rate of 5 gal/min, emptying into tank B at the same rate. In turn, salt solution leaves tank B at 5 gal/min, emptying into tank C at the same rate. Finally, salt solution leaves tank C at 5 gal/min. Because of the constant flow to and from each tank, the volume of solution in each tank remains constant for all time. Tank A initially contains 15 pounds of salt. Use `odesolve` to estimate the times when the salt content reaches a maximum in tanks B and C.

6. A driven RC-circuit is modeled with the initial value problem

$$V_C' = -V_C + \cos(2\pi f t), \qquad V_C(0) = 0,$$

where f is the frequency of the driving force. For purposes of this demonstration, think of the driving force $V_I = \cos(2\pi f t)$ as the *input signal* and V_C, the voltage response across the capacitor, as the *output signal*. This circuit is called a *low-pass filter*. The goal of this problem is to vary the frequency of the input signal and measure the response of the circuit by directly comparing the output and input signals. We can get `odesolve` to do this for us with a little subterfuge. If we differentiate the input signal we get $V_I' = -2\pi f \sin(2\pi f t)$. We add this equation to the equation for the circuit to create the system

$$V_C' = -V_C + V_I,$$
$$V_I' = -2\pi f \sin(2\pi f t).$$

Note that we want $V_I(0) = \cos(2\pi f(0)) = 1$. We will have `odesolve` solve the system, and plot the input and output signals on the same figure, allowing easy comparison.

a) Set the frequency to $f = 0.04$ Hz. Recall that the period and frequency are related by $T = 1/f$, so the period in this case is $T = 25$ seconds. Enter the system in `odesolve`, set the initial conditions to $V_C(0) = 0$ and $V_I(0) = 1$, and enter the parameter $f = 0.04$. We want to solve over five periods, so set the solution interval to $[0, 5T] = [0, 5/f]$. Solve and plot both V_C and V_I. How severely, if at all, is the output signal attenuated (reduced in amplitude) in comparison to the input signal?

b) Gradually increase the frequency of the input signal from $f = 0.04$ Hz to $f = 1$ Hz, adjusting the solution interval in each case to $[0, 5T] = [0, 5/f]$, graphing the input and output signal over five periods of the input signal.

c) Describe what happens to the amplitude of the output signal in comparison to the input signal as you increase the frequency of the input signal V_I? Why is this circuit called a low-pass filter?

7. A driven LRC-circuit is modelled by the equation $LQ'' + RQ' + Q/C = E$, where Q is the charge on the capacitor, L is the inductance, R the resistance, C the capacitance, and E is the driving term, our input voltage in this case. We consider E to be the input signal, and we will measure the output by measuring Q. We will use `odesolve` to compare the input and output graphically.

By introducing the current $I = Q'$, the second order differential equation becomes a system of two first order equations, $Q' = I$, $I' = -RI/L - Q/(LC) + E/L$. We will use the complicated input $E = -\cos t - (4/5)\cos 5t$. Differentiating E we get the three dimensional system

$$Q' = I$$

$$I' = -\frac{R}{L}I - \frac{1}{LC}Q + \frac{1}{L}E$$

$$E' = \sin t + 4\sin 5t.$$

Set the initial charge and current equal to zero, so $Q(0) = 0$ and $I(0) = 0$. Notice that $E(0) = -9/5$.

a) Enter the system and the initial conditions into odesolve. Enter the parameters $L = 1\,\text{H}$, $R = 0.1\,\Omega$, and $C = 0.01$. Solve and plot the charge Q and source voltage E versus time t on the time interval $[0, 100]$.

b) We are going to show how the circuit can be tuned by changing the capacitance C to isolate the two components of our complicated input signal $E = -\cos t - (4/5)\cos 5t$. If the input signal has the simpler form $E = A\cos \omega t$, one can show that that the amplitude of the charge on the capacitor reaches a maximum when

$$\omega = \sqrt{\frac{1}{LC} - \frac{R^2}{2L^2}}. \tag{9.4}$$

Use (9.4) to calculate the value of C that will tune the circuit to extract (and perhaps amplify) the $\cos t$ of the input signal (note that $\omega = 1$ here). Use odesolve to plot Q and E on the time interval $[0, 100]$, which is long enough for transients to die out. Then use the "Zoom in" tool on the "Tools" menu to crop the image to the interval $[75, 100]$. Explain how you know that the $\cos t$ component of the input signal is being isolated in the output.

c) Follow the procedure in part b) to find the value of C that will isolate the $\cos 5t$ component of the input signal. Again, explain how you know that this is the component being picked up.

8. Something called a *scroll circuit* is a double looped circuit with two capacitors, an inductor, and a nonlinear resistor. After scaling, the circuit is described by the following system of equations,

$$x' = -cf(y - x)$$
$$y' = -f(y - x) - z$$
$$z' = ky$$

where

$$f(v) = -av + 0.5(a + b)(|v + 1| - |v - 1|)$$

describes the current passing through the nonlinear resistor as a function of the voltage drop across the resistor (after scaling).

a) Enter the equations in odesolve as follows.

```
x' = -c*p
y' = -p - z
z' = k*y
```

b) Enter the initial conditions $x(0) = y(0) = z(0) = 1$ and the parameters $a = 0.07$, $b = 0.1$, $c = 0.1$, and $k = 1$.

c) Enter the following in an expression edit field.

```
p = -a*(y - x) + 0.5*(a + b)*(abs(y - x + 1) - abs(y - x - 1))
```

d) Set the relative tolerance to 1e-6. Obtain plots of the solution in xyz-space and the projection on the xz-plane. Repeat the experiment for $c = 33$.

9. The system

$$x_1' = 0.002x_1$$
$$x_2' = 0.002x_1 - 0.08x_2 - x_2x_3^2$$
$$x_3' = 0.08x_2 - x_3 + x_2x_3^2$$
$$x_4' = x_3$$

137

models a chemical reaction during which two components, x_2 and x_3, undergo oscillatory growth and decay. Enter the equations in odesolve, then set initial conditions $x_1(0) = 500$ and $x_2(0) = x_3(0) = x_4(0) = 0$.

a) Set the relative tolerance to 1e-4 and plot x_2 and x_3 versus time.

b) If you attempt to plot all four components versus time, they scale badly in the odesolve window. Try it! Export the solution data to the command window workspace. Take careful notice of the name of the structure that is exported. If you have not exported anything during this session of MATLAB it will be odedata1. We will assume that is the case, but the output variable could be odedata* where * is a small integer. Enter odedata1 at the MATLAB prompt to see the form of the structure and a list of its fields. Assuming that the fields are t, x1, x2, x3, and x4, set x1 = odedata1.x1/200; x2 = odedata1.x2; x3 = odedata1.x3; x4 = odedata1.x4/200; and t = odedata1.t;. Notice that we are rescaling x1 and x4. Plot x1, x2, x3, and x4 versus time in a single figure. Label your axes and provide appropriate labels, title, and legend.

10. The *Rossler System*

$$x' = -y - z$$
$$y' = x + ay$$
$$z' = b - cz + xz$$

provides another example of a *strange attractor* similar to that of the Lorenz System.

a) Set the parameters to $a = 0.2, b = 0.2, c = 2.5$, and use the initial conditions $x(0) = y(0) = z(0) = 1$. Set the maximum step size of the solver to 0.5. Select 3D plot and select the variables x, y, and z. solve and plot the solution on the time interval [0, 250]. Export the data to the command window workspace as the structure odedata*, where 8 is a small integer. Then execute t = odedata*.t, x = odedata*.x, y = odedata*.y, and z = odedata*.z. Eliminate transient behavior with N=find(t>200), followed by plot(x(N),y(N) to plot in the xy-plane.

b) Repeat the instructions of part a) for $c = 3$, $c = 4$, and $c = 4.2$. Obtain xy-plots for each c-value. Be sure to eliminate transient behavior as was done in part a).

c) Set $c = 5$ and and the time interval to [0, 1000]. Repeat part a), but restrict the xy-plot to the time interval [500, 1000].

d) The progression in parts a) and b) is called *period doubling* on the route to chaos, which is illustrated in part c). Can you explain why?

138

10.   The Symbolic Toolbox

The Symbolic Toolbox provides a link between MATLAB and the symbolic algebra program known as Maple. Unlike the rest of MATLAB, which provides powerful numerical routines, the Symbolic Toolbox deals with symbols, formulas, and equations. In dealing with differential equations, it leads to explicit formulas for the solutions, provided such formulas exist at all.

The Symbolic Toolbox is included with the Student Edition of MATLAB, however, it is distributed as an additional toolbox in the standard version of MATLAB. Therefore it may not be available on your computer. If you are unsure whether or not it is available, execute the command `help symbolic` at the MATLAB prompt. If the Symbolic Toolbox is available, you will get a list of all of the commands available in the toolbox. If it is not, you will get an error message, saying `symbolic not found`.

In this chapter, we will put our emphasis on using the Symbolic Toolbox to solve differential equations. This means that MATLAB will try to tell you the exact, analytic solution to an equation, or even to an initial value problem. MATLAB can be successful, of course, only if such a solution exists. To learn more about the Symbolic Toolbox select **Help→MATLAB Help** from the command window menu, and look for the *Symbolic Math Toolbox*.

Finally, a disclaimer. We are using version 3.0.1 of the Symbolic Toolbox, the latest version available at the time of this writing. The Symbolic Toolbox has gone through a number of changes recently. If you are not using this version you may notice some differences in the responses to commands. In fact we have noticed different responses from version 3.0.1 operating on different computers. None of the differences should be serious, however.

Algebraic Expressions and Symbolic Objects

We have already seen at least three classes of objects in MATLAB. Any numerical quantity, be it a number or a matrix, is an object of the class `double`. In addition, we have used strings, which are objects of class `char`, and function handles, which are objects in the class `function_handle.`. The Symbolic Toolbox uses algebraic expressions, which are objects of class `sym`.[1] For example, the commands

```
>> clear all, x = 5; str = 'hello world'; ods = @ode45; y = sym('alpha');
```

will clear all previously defined variables, and then create four new objects. The object x is an array of class `double` with value 5, `str` is a string of class `char`, `ods` is a function handle, and y is an object of class `sym` with value `alpha`. The `whos` command reveals the class of objects in your workspace.

```
>> whos
  Name      Size        Bytes  Class

  ods       1x1            16  function_handle array
  str       1x11           22  char array
```

---

[1] There are several more classes defined within MATLAB. To see a partial list, execute `help class`. It is even possible to define your own. For a detailed account of object-oriented programming see *Classes and Objects* in the online MATLAB documentation.

```
      x              1x1                    8  double array
      y              1x1                  134  sym object
```

The last column of the output from the `whos` command lists the class of each object in your workspace.

You can examine the contents of objects in your workspace in the usual manner. For example,

```
>> x,y,ods,str
x =
     5
y =
alpha
ods =
     @ode45
str =
hello world
```

There is a subtle difference in ways objects of the various classes are displayed in the command window. The contents of the `sym` and `char` objects are left justified, while the contents of objects of class `double` and `function_handle` are indented by slightly different amounts. The class of an object can also be found by using the `class` command. For example, `class(y)` will reveal that y is an object of class `sym`.

In the previous example, the command `sym` was used to create an instance of class `sym`. However, you can also use the `sym` command to change the class of an existing object. For example, the command

```
>> z = 5.7
z =
     5.7000
```

creates an object z of class `double`, which is readily checked with the command

```
>> class(z)
ans =
double
```

You can change the class of z to `sym` with the command

```
>> z = sym(z)
z =
57/10
```

Check the class of z by typing `class(z)` at the MATLAB prompt. Note the left justification of the `sym` object z. Notice also that MATLAB has changed to rational notation for z.

Differentiation. Some MATLAB functions will work on only one class of objects. However, many will work on more than one class, often with completely different results. This technique is known as *overloading* the function. For example, if you execute `help diff`, you will discover that `diff` operates on a vector of class `double` by taking differences of consecutive elements of the vector. For example:

```
>> x = [1 2 4 7 11]
x =
      1     2     4     7    11
>> diff(x)
ans =
      1     2     3     4
```

The function `diff` has been overloaded to differentiate objects of class `sym`. For example:

```
>> g = sym('x^2')
g =
x^2
>> diff(g)
ans =
2*x
```

In this case, MATLAB recognizes that the input argument has class `sym` and promptly takes the derivative of $g = x^2$ with respect to x. You will have noticed that the `help diff` command gives no hint of this new power of the function `diff` until the last few lines, where it lists the overloaded methods. To see a description of this overloaded method it is necessary to indicate that you want to see how `diff` works on objects of class `sym`. This is done with the command `help sym/diff`. Try it!

You can declare symbolic objects one at a time, such as

```
>> x = sym('x'); a = sym('a'); b = sym('b'); c = sym('c');
```

or you can use the `syms` command to declare them all at once, as in

```
>> syms x a b c
```

The order of declaration is unimportant.

The operators +, −, *, /, and ^ have overloaded definitions for use on symbolic objects, so you can use them to create algebraic expressions. For example, since a, b, c, and x have been declared to be symbolic objects,

```
>> f = a^2*x^2 + b*x + c;
```

creates a new symbolic object `f`. We can compute the derivative of `f` with the command

```
>> diff(f)
ans =
2*a^2*x+b
```

We can just as easily take a second derivative.

```
>> diff(f,2)
ans =
2*a^2
```

The default variable. How did the Symbolic Toolbox know to differentiate with respect to x, and not with respect to another symbolic variable, say a? The answer lies in the Symbolic Toolbox's *default variable*:

The default symbolic variable in a symbolic expression is the variable that is closest to 'x' alphabetically.

The findsym command will find all of the symbolic variables in a symbolic expression.

```
>> findsym(f)
ans =
a, b, c, x
```

The command findsym(f,n) returns the n symbolic variables which are closest to x in the alphabet. Hence,

```
>> findsym(f,1)
ans =
x
```

returns the default symbolic variable. That is why the Symbolic Toolbox differentiated with respect to x in the above example. Of course, most Symbolic Toolbox commands allow the user to override the default symbolic variable. For example,

```
>> diff(f,a)
ans =
2*a*x^2
```

finds the derivative of f with respect to a, and

```
>> diff(f,a,2)
ans =
2*x^2
```

finds the second derivative of f with respect to a.

Verifying Solutions to Differential Equations

After having completed extensive paper and pencil work to obtain the solution of a differential equation or initial value problem, you might want to check your answer. Why not let the Symbolic Toolbox do the work for you?

Example 1. *Verify that* $y(t) = ce^{-t^2}$ *is a solution of*

$$y' + 2ty = 0, \tag{10.1}$$

where c is an arbitrary constant.

142

Enter

```
>> syms y t c
>> y = c*exp(-t^2);
```

to define the symbolic variables and the potential solution. Then

```
>> diff(y,t) + 2*t*y
ans =
0
```

verifies that you have a solution.[2]

Example 2. *Verify that* $y = x \tan x$ *is a solution of*

$$xy' = y + x^2 + y^2. \qquad (10.2)$$

Note that the independent variable is x. Enter

```
>> syms x y
>> y = x*tan(x);
```

to define the symbolic variables and enter the potential solution. You can enter each side of equation (10.2) and compare them, but it is usually easier to enter their difference and verify that it equals zero. In our case

```
>> x*diff(y,x) - (y+x^2+y^2)
ans =
x*(tan(x)+x*(1+tan(x)^2))-x*tan(x)-x^2-x^2*tan(x)^2
```

does not seem to be equal to 0. However, this is a complicated expression and it is possible that if we are clever enough we might be able to simplify it. Let's let MATLAB do it for us using the Symbolic Toolbox's simple command. Recall that ans always contains the answer to the last computation. Then

```
>> zzz = simple(ans)
zzz =
0
```

verifies that $y = x \tan x$ is a solution of equation (10.2).

We have hidden a lot of detail by using the command simple in the way we did. The command simple causes MATLAB to cycle through a large number of simplification algorithms in search of the one that yields the output that is shortest in length. Forget the zzz and execute simple(ans). Without an output variable simple shows the output of each of these algorithms, demonstrating how powerful simple really is.

---

[2] We originally used C as the constant in the definition of y. However, there seems to be a minor bug in version 3.0.1 of the Symbolic Toolbox with the use of the capital letter C. Avoid using this letter unless MATLAB introduces it.

Solving Ordinary Differential Equations

The routine `dsolve` is probably the most useful differential equations tool in the Symbolic Toolbox. To get a quick description of this routine, enter `help dsolve`. The best way to learn the syntax of `dsolve` is to look at an example.

Example 3. *Find the general solution of the first order differential equation*

$$\frac{dy}{dt} + y = te^t. \tag{10.3}$$

The Symbolic Toolbox easily finds the general solution of the differential equation in (10.3) with the command

```
>> dsolve('Dy + y = t*exp(t)')
ans =
1/2*t*exp(t)-1/4*exp(t)+exp(-t)*C1
```

Notice that `dsolve` requires that the differential equation be entered as a string, delimited with single quotes. The notation Dy is used for y', D2y for y'', etc.

The `dsolve` routine assumes that the independent variable is t. You can override this default by placing the independent variable of choice as the last input argument to `dsolve`. For example, `dsolve('Dy + y = x*exp(x)','x')` declares that the independent variable is x. Try this and then try `dsolve('Dy + y = x*exp(x)')` to see the difference. It is a good practice to always declare the independent variable when using the `dsolve` command.

Note that the output was not assigned to the variable y, as one might expect. This is easily rectified.

```
>> y = dsolve('Dy + y = t*exp(t)','t')
y =
1/2*t*exp(t)-1/4*exp(t)+exp(-t)*C1
```

The command

```
>> pretty(y)
```

$$1/2 \ t \ \exp(t) \ - \ 1/4 \ \exp(t) \ + \ \exp(-t) \ C1$$

gives us a nicer display of the output.

Substituting values and plotting. Let's substitute a number for the constant C1 and obtain a plot of the result on the time interval $[-1, 3]$. The commands

```
>> C1 = 2
C1 =
      2
>> y = subs(y)
y =
1/2*t*exp(t)-1/4*exp(t)+2*exp(-t)
```

144

replace the variable C1 in the symbolic expression y with the value of C1 in MATLAB's workspace.[3]

MATLAB's ezplot command has been overloaded for symbolic objects, and the command

```
>> ezplot(y,[-1,3])
```

will produce an image similar to that in Figure 10.1. Execute `help sym/ezplot` to learn about the many options available in `ezplot`.

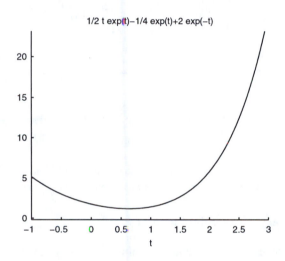

Figure 10.1. The plot of $(1/2)te^t - (1/4)e^t + 2e^{-t}$ on $[-1, 3]$.

Example 4. *Consider the equation of an undamped oscillator with periodic forcing function*

$$y'' + 144y = \cos 11t, \tag{10.4}$$

with initial position given by $y(0) = 0$ and initial velocity given by $y'(0) = 0$. Plot a solution of the displacement y with these given initial conditions.

The Symbolic Toolbox easily finds solutions of initial value problems. In our case, enter

```
>> y = dsolve('D2y + 144*y = cos(11*t)','y(0) = 0,Dy(0) = 0','t')
y =
-1/23*cos(12*t)+1/23*cos(11*t)
```

We can plot this solution on the time interval $[0, 6\pi]$ with the command

```
>> ezplot(y,[0,6*pi])
```

---

[3] MATLAB's `subs` command is a multi-faceted, complex command. Execute `help subs` to learn more. Notice that the command `y = subs(y)` makes a substitution for every variable which has a value in MATLAB's workspace. If we are not careful, this can lead to serious mistakes.

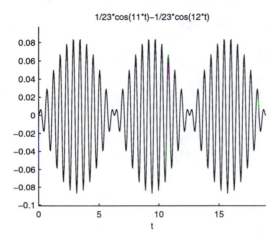

Figure 10.2. The solution of $y'' + 144y = \cos 11t$, $y(0) = 0$, $y'(0) = 0$.

producing an image similar to that in Figure 10.2. The phenomenon displayed in Figure 10.2 is called *beats*, a condition present in undamped, forced oscillators that are nearly resonant.

Solving Systems of Ordinary Differential Equations

The routine dsolve solves systems of equations almost as easily as it does single equations, as long as there actually is an analytical solution.

Example 5. *Find the solution of the initial value problem*

$$x' = -2x - 3y,$$
$$y' = 3x - 2y,$$

(10.5)

with $x(0) = 1$ and $y(0) = -1$.

It is a good idea to clear variables from your workspace when they are no longer needed. Left-over variables can wreak unexpected havoc when employing the subs command. You can delete all[4] variables from your workspace with

```
>> clear all
```

Enter the whos command to verify that there are no variables left in the workspace.

We can call dsolve in the usual manner. Readers will note that the input to dsolve in this example parallels that in Examples 3 and 4. The difference lies in the output variable(s). If a single variable is used

---

[4] You can also select individual variables for deletion. For example, the command clear t clears only the variable t from your workspace.

to gather the output, then the Symbolic Toolbox returns a MATLAB structure[5]. A *structure* is an object which has user defined fields, each of which is a MATLAB object. In this case the fields are symbolic objects containing the components of the solution[6] of system (10.5). Thus the comand

```
>> S = dsolve('Dx = -2*x - 3*y,Dy = 3*x-2*y', 'x(0) = 1,y(0) = -1','t')
S =
    x: [1x1 sym]
    y: [1x1 sym]
```

produces the structure S with fields

```
>> S.x
ans =
exp(-2*t)*(cos(3*t)+sin(3*t))
>> S.y
ans =
exp(-2*t)*(sin(3*t)-cos(3*t))
```

It is also possible to output the components of the solution directly with the command

```
>> [x,y] = dsolve('Dx = -2*x-3*y,Dy = 3*x-2*y','x(0) = 1, y(0) = -1','t')
x =
exp(-2*t)*(cos(3*t)+sin(3*t))
y =
exp(-2*t)*(sin(3*t)-cos(3*t))
```

Each component of the solution can be plotted versus t with the following sequence of commands.

```
>> ezplot(x,[0,2*pi])
>> hold on
>> ezplot(y,[0,2*pi])
>> legend('x','y')
>> hold off
```

Adjust the view with the `axis` command, and add a title and axis labels.

```
>> axis([0,2*pi,-1.2,1.2])
>> title('Solutions of x'' = -2x-3y, y'' = 3x-2y, x(0) = 1, y(0) = -1.')
>> xlabel('Time t')
>> ylabel('x and y')
```

This should produce an image similar to that in Figure 10.3. We used the Property Editor to make one of the curves dashed.

It is also possible to produce a phase-plane plot of the solution. With the structure output the basic command is simply `ezplot(S.x,S.y)`. After some editing we get Figure 10.4.

---

[5] A structure is an object of the class `struct`, another of the many MATLAB classes. For a complete introduction to structures and cell arrays, see the online MATLAB documentation.

[6] The authors got different but equivalent solutions on different computers.

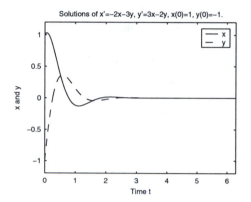

Solutions of x'=-2x-3y, y'=3x-2y, x(0)=1, y(0)=-1.

Figure 10.3. Plots of x and y versus t.

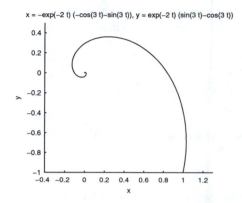

$x = -\exp(-2t)(-\cos(3t)-\sin(3t))$, $y = \exp(-2t)(\sin(3t)-\cos(3t))$

Figure 10.4. The phase-plane plot of y versus x.

Interpreting the Output of DSOLVE

Dsolve can output some strange looking solutions in certain situations.

Example 6. *Find the solution of the differential equation*

$$2tyy' = 3y^2 - t^2. \tag{10.6}$$

Note that equation (10.6) is nonlinear. Dsolve will find exact solutions of some nonlinear differential equations.

```
>> y = dsolve('2*t*y*Dy = 3*y^2 - t^2','t')
y =
[   (1+t*C1)^(1/2)*t]
[ -(1+t*C1)^(1/2)*t]
```

The output y is a vector. You can access each component of this vector in the usual manner; that is, y(1) is the first component, and y(2) is the second component. Try it!

Looking at y closely, we see that MATLAB is saying that the general solution to (10.6) is $y(t) = \pm t\sqrt{1 + C_1 t}$. When solving an initial value problem, we need to use the initial condition to decide which sign to use as well as to determine the constant C_1.

Sometimes the Symbolic Toolbox cannot find explicit solutions (where the dependent variable is expressed explicitly in terms of the independent variable).

Example 7. *Find the solution of the initial value problem*

$$(1 - \sin x)x' = t, \quad x(0) = 0. \tag{10.7}$$

148

Equation (10.7) is separable. The general solution is given by the implicit formula

$$x + \cos x = \frac{t^2}{2} + 1. \tag{10.8}$$

Of course, it is not possible to solve equation (10.8) for x in terms of t. When we execute

```
>> dsolve('(1 - sin(x))*Dx = t', 'x(0) = 0','t')
```

to see what MATLAB will do, we get an error message saying that MATLAB will not look for implicit solutions when initial conditions are given. So we execute

```
>> dsolve('(1 - sin(x))*Dx = t','t')
Warning: Explicit solution could not be found; implicit solution returned.
> In C:\MATLAB6p5\toolbox\symbolic\dsolve.m at line 292
ans =
1/2*t^2-x-cos(x)+C1 = 0
```

This time, MATLAB tells us that it could not find an explicit solution and gives us the formula for the general implicit solution. We can find out what the constant has to be by using the initial conditions and the subs command.

```
>> x = 0; t = 0;
>> subs(ans)
ans =
-1+C1 = 0
```

Thus, C1 = 1 and we see that the solution to the initial value problem satisfies the implicit equation $(1/2)t^2 - x - \cos x + 1 = 0$, which is equivlalent to (10.8).[7]

The Solve Command

The solve command is the Symbolic Toolbox's versatile equation solver. It performs equally well with single equations and multiple equations. For example, if you need to solve the equation $ax + b = 0$, then the command

```
>> solve('a*x + b')
ans =
-b/a
```

provides the solution, $x = -b/a$. We need to make two important points about the syntax of the solve command. First, the command solve('a*x+b') is identical to the command solve('a*x+b=0'). Try

---

[7] Different versions of the Symbolic Toolbox give quite different responses when implicitly defined solutions appear.

it! Second, we did not indicate what variable to solve for, so the `solve` command found a solution for the default variable.[8] You can choose a different solution variable. The command

```
>> solve('a*x+b','a')
ans =
-b/x
```

solves the equation $ax + b = 0$ for a.

If an equation has multiple solutions, they are listed in an output vector.

```
>> s = solve('a*x^2 + b*x + c')
s =
[ 1/2/a*(-b+(b^2-4*a*c)^(1/2))]
[ 1/2/a*(-b-(b^2-4*a*c)^(1/2))]
```

You can access the components of the solution vector in the usual manner.

The command

```
>> pretty(s(1))
```

$$
1/2 \ \frac{-b + (b^2 - 4\ a\ c)^{1/2}}{a}
$$

will display the formula in more understandable form.

Finally, if the Symbolic Toolbox cannot find a symbolic solution, it will attempt to find a numerical solution.

```
>> s = solve('x + 2 = exp(x^2)')
s =
1.0571035499947380777922377221533
```

That's 32 significant figures. If you want more, say 40, execute

```
>> digits(40)
>> s = solve('x+2 = exp(x^2)')
s =
1.057103549994738077792237722153280429451
```

Example 8. *The logistic model for population growth is given by the equation*

$$
\frac{dp}{dt} = k\left(1 - \frac{P}{N}\right)P, \tag{10.9}
$$

where k is the natural growth rate of the population and N is the carrying capacity of the environment. Assume that $k = 0.025$ and $N = 1000$. How long will it take an initial population of 200 to grow to 800? Assume that time is measured in minutes.

---

[8] Type `findsym(sym('a*x+b'),1)` to find the default variable in the expression $ax + b$.

150

With $k = 0.025$ and $N = 1000$, equation (10.9) becomes $dP/dt = 0.025(1 - P/1000)P$, with initial population given by $P(0) = 200$. Using dsolve,

```
>> P = dsolve('DP = 0.025*(1-P/1000)*P','P(0) = 200','t')
P =
1000/(1+4*exp(-1/40*t))
```

we find that the solution of the initial value problem is

$$P = \frac{1000}{1 + 4e^{-t/40}}. \tag{10.10}$$

The following commands provide a plot similar to that in Figure 10.5.

```
>> ezplot(P,[0,300])
>> grid
>> xlabel('Time in minutes')
>> ylabel('Population')
>> title('A Logistic Model')
```

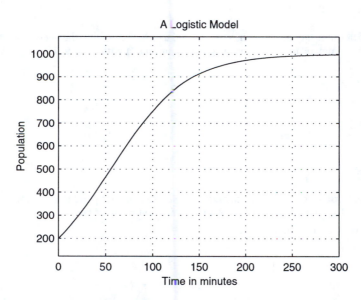

Figure 10.5. The solution of $dP/dt = 0.025(1 - P/1000)P$, $P(0) = 200$.

If you examine the plot in Figure 10.5, it appears that the population reaches a level of 800 somewhere between 100 and 150 minutes. Substituting $P = 800$ in equation (10.10), we see that we want to find t so that

$$800 = \frac{1000}{1 + 4e^{-t/40}}.$$

For this we can use the commands[9]

```
>> s = solve('800 = 1000/(1+4*exp(-t/40))','t')
s =
40*log(16)
>> double(s)
ans =
   110.9035
```

Consequently, it takes approximately 111 minutes for the population to reach 800.

Exercises

In Exercises 1 – 5, use MATLAB and the technique demonstrated in Examples 1 and 2 to verify that y is a solution to the indicated equation.

1. $y = 1 + e^{-t^2/2}$, $y' + ty = t$.
2. $w = 1/(s - 3)$, $w' + w^2 = 0$.
3. $y = 10 - t^2/2$, $yy' + ty = 0$.
4. $y = \ln(x)$, $(2e^y - x)y' = 1$. Recall: $\ln(x)$ is $\log(x)$ in MATLAB.
5. $y = e^t \cos(2t)$, $y'' - 2y' + 5y = 0$.

In Exercises 6 – 11, determine the independent variable, and use `dsolve` to find the general solution to the indicated equation. Use the `subs` command to replace the integration constant C1 with C1 = 2. Use `ezplot` to plot the resulting solution.

6. $y' + ty = t$.
7. $y' + y^2 = 0$.
8. $yy' + ty = 0$.
9. $(2e^y - x)y' = 1$. Hint: Use `dsolve('(2*exp(y)-x)*Dy = 1','x')`.
10. $(x + y^2)y' = y$.
11. $x(y' - y) = e^x$.

In Exercises 12 – 17, determine the independent variable, and use `dsolve` to find the solution to the indicated initial value problem. Use `ezplot` to plot the solution over the indicated time interval.

12. $y' + ty = t$, $y(0) = -1$, $[-4, 4]$.
13. $y' + y^2 = 0$, $y(0) = 2$, $[0, 5]$.
14. $yy' + ty = 0$, $y(1) = 4$, $[-4, 4]$.
15. $(2e^y - x)y' = 1$, $y(0) = 0$, $[-5, 5]$. Hint: `dsolve('(2*exp(y)-x)*Dy=1','y(0)=0','x')`.
16. $(x + y^2)y' = y$, $y(0) = 4$, $[-4, 6]$.
17. $x(y' - y) = e^x$, $y(1) = 4e$, $[0.0001, 0.001]$, and $[0.001, 1]$.

In Exercises 18 and 19, use `dsolve` to obtain the solution of each of the indicated second order differential equations. Use the `simple` command to find the simplest form of that solution. Use `ezplot` to sketch the solution on the indicated time interval.

---

[9] You can actually get pretty fancy here and enter `solve(P-800)`. This is pretty sophisticated use of the Symbolic Toolbox, and indicates some of the power you can attain with experience.

18. $y'' + 4y = 3\cos 2.1t$
    $y(0) = 0,\ y'(0) = 0,\ [0, 64\pi]$

19. $y'' + 16y = 3\sin 4t$
    $y(0) = 0,\ y'(0) = 0,\ [0, 32\pi]$

20. Use `dsolve` to obtain the solution of $y'' + y' + 144y = \cos 2t$, $y(0) = 0$, $y'(0) = 0$. Use `ezplot` to plot the solution on the time interval $[0, 6\pi]$. Use `ode45` to find a numerical solution on the same time interval and superimpose that plot of the first plot. How closely does the numerical solution match the symbolic solution?

21. Use the technique of Examples 1 and 2 to verify that

$$x = e^{-t}(\cos 2t - \sin 2t),$$
$$y = e^{-t}(3\sin 2t - \cos 2t),$$

is a solution of the system

$$x' = -5x - 2y,$$
$$y' = 10x + 3y.$$

Furthermore, use the `subs` command to verify that the solution satisfies the initial conditions $x(0) = 1$ and $y(0) = -1$.

22. Use `dsolve` to solve the initial value problem

$$y' = v,$$
$$v' = -2v - 2y + \cos 2t,$$

with $y(0) = 1$ and $v(0) = -1$, over the time interval $[0, 30]$.

In Exercises 23 – 25, find the solutions to the indicated initial value problem.

23. $t^2 y'' = (y')^2$, with $y(1) = 3$ and $y'(1) = 2$.

24. $y'' = yy'$, with $y(0) = 0$, and $y'(0) = 2$.

25. $y'' = ty' + y + 1$, with $y(0) = 1$, and $y'(0) = 0$.

26. Suppose we start with a population of 100 individuals at time $t = 0$, and that the population is correctly modeled by the logistic equation. Suppose that at time $t = 2$ there are 200 individuals in the population, and that the population approaches steady state at a population of 1000. Plot the population over the interval $[0, 20]$. What is the population at time $t = 10$?

27. Suppose we have a population which is correctly modeled by the logistic equation, and experimental measurements show that $p(0) = 50$, $p(1) = 150$, and $p(2) = 250$. Use the Symbolic Toolbox to derive the formula for the population as a function of time. Plot the population over the interval $[0, 5]$. What is the limiting value of the population? What is $p(3)$?

11.   Linear Algebra Using MATLAB

MATLAB is short for "Matrix Laboratory." It contains a large number of built-in routines that make matrix algebra very easy. In this chapter we will learn how to use some of those routines to solve systems of linear equations. In the next chapter we will use the linear algebra learned here to help us solve systems of differential equations.

Systems of Linear Equations

Let's start by looking at a very elementary example.

Example 1. *Find the solution to the system*

$$\begin{aligned} u + 2v &= 5, \\ 4u - v &= 2. \end{aligned} \tag{11.1}$$

Using the method usually taught in secondary school, we solve the first equation for the variable u, yielding $u = 5 - 2v$, and substitute this result into the second equation. In this way we obtain a new system consisting of the first equation and the modified second equation

$$\begin{aligned} u + 2v &= 5, \\ -9v &= -18, \end{aligned} \tag{11.2}$$

which has the same solutions as (11.1). The advantage is that the variable u has been *eliminated* from the second equation. Therefore, system (11.2) is quite easily solved by the method known as *backsolving*. This means that we solve the system by starting with the last equation and move up. From the second equation in (11.2), we see that $v = 2$. Substituting $v = 2$ into the first equation, we can solve for u, getting $u = 1$.

The system in (11.1) is a system of two equations with two unknowns. We want to find a systematic method that will work in general for m equations involving n unknowns, where m, and n are possibly very large, and not necessarily equal. Our goal is to find *all* solutions of such a system. The general method is not significantly different from the method of elimination and backsolving used in Example 1.

Linear Combinations of Vectors and the Matrix-Vector Product

We need to develop some algebraic ideas to facilitate our solution method. We can put both sides of equation (11.1) into vectors, so that it reads

$$\begin{bmatrix} u + 2v \\ 4u - v \end{bmatrix} = \begin{bmatrix} 5 \\ 2 \end{bmatrix}. \tag{11.3}$$

Using standard arithmetic operations with vectors, we can rewrite the left-hand side so that (11.3) becomes

$$u \begin{bmatrix} 1 \\ 4 \end{bmatrix} + v \begin{bmatrix} 2 \\ -1 \end{bmatrix} = \begin{bmatrix} 5 \\ 2 \end{bmatrix}. \tag{11.4}$$

The form of the left-hand side in (11.4) is so important we give it a name. It is called the *linear combination* of the vectors $[1, 4]^T$ and $[2, -1]^T$ with coefficients u and v. In general, let $\mathbf{v}_1, \mathbf{v}_2, \ldots, \mathbf{v}_p$ be vectors of the same size, and let $x_1, x_2, \ldots, x_p$ be scalars.[1] Then

$$x_1 \mathbf{v}_1 + x_2 \mathbf{v}_2 + \cdots + x_p \mathbf{v}_p$$

is called the linear combination of the vectors $\mathbf{v}_1, \mathbf{v}_2, \ldots, \mathbf{v}_p$ with coefficients $x_1, x_2, \ldots, x_p$. With this new terminology, the problem of solving the system in (11.1) becomes that of finding the coefficients u and v so that the linear combination with these coefficients is equal to the right-hand side in (11.4).

The concept of linear combination, together with its importance in understanding systems of equations as indicated in equation (11.4), leads us to the definition of *matrix-vector multiplication*. We formally define the product of a matrix A and a vector $\mathbf{x}$ to be the vector $A\mathbf{x}$ which is the linear combination of the columns of A, using the entries of the vector $\mathbf{x}$ for the scalars in the linear combination. Thus, if A is the matrix which has the vectors $\mathbf{v}_1, \mathbf{v}_2, \ldots, \mathbf{v}_p$ as columns and $\mathbf{x} = [x_1, x_2, \ldots, x_p]^T$, then

$$A\mathbf{x} = x_1 \mathbf{v}_1 + x_2 \mathbf{v}_2 + \cdots + x_p \mathbf{v}_p.$$

Let's return to the system in (11.1). We reinterpreted the system as a linear combination in (11.4). In view of the definition of matrix-vector product, we can write (11.4) as

$$\begin{bmatrix} 1 & 2 \\ 4 & -1 \end{bmatrix} \begin{bmatrix} u \\ v \end{bmatrix} = \begin{bmatrix} 5 \\ 2 \end{bmatrix}. \tag{11.5}$$

If we define

$$A = \begin{bmatrix} 1 & 2 \\ 4 & -1 \end{bmatrix}, \quad \mathbf{x} = \begin{bmatrix} u \\ v \end{bmatrix}, \quad \text{and} \quad \mathbf{b} = \begin{bmatrix} 5 \\ 2 \end{bmatrix}, \tag{11.6}$$

the system of equations can be written as $A\mathbf{x} = \mathbf{b}$. If we compare equations (11.5) and (11.6) to the original system in (11.1) we see that the matrix A contains the coefficients of the system. It is therefore called the *coefficient matrix*. Similarly, the vector $\mathbf{b}$ is called the *right-hand side* and the vector $\mathbf{x}$ is called the *vector of unknowns* or the *unknown vector*.

Let's look at the concepts of linear combination and matrix multiplication for a more complicated example.

Example 2. *Consider the vectors*

$$\mathbf{a}_1 = \begin{bmatrix} 2 \\ -2 \\ -2 \end{bmatrix}, \quad \mathbf{a}_2 = \begin{bmatrix} 4 \\ 3 \\ 4 \end{bmatrix}, \quad \text{and} \quad \mathbf{a}_3 = \begin{bmatrix} -3 \\ 2 \\ 1 \end{bmatrix}.$$

Form and simplify the linear combination $1\mathbf{a}_1 + 0\mathbf{a}_2 - 2\mathbf{a}_3$.

---

[1] Scalars are simply numbers. In most examples here they are real numbers, but they could also be complex numbers. Indeed, the vectors themselves could have complex entries.

This is easily accomplished in MATLAB.

```
>> a1 = [2;-2;-2]; a2 = [4;3;4]; a3 = [-3;2;1];
>> 1*a1 + 0*a2 - 2*a3
ans =
     8
    -6
    -4
```

Example 3. *Find the matrix-vector product A*$\mathbf{x}$*, where*

$$A = \begin{bmatrix} 2 & 4 & -3 \\ -2 & 3 & 2 \\ -2 & 4 & 1 \end{bmatrix} \quad and \quad \mathbf{x} = \begin{bmatrix} 1 \\ 0 \\ -2 \end{bmatrix}.$$

Note that the columns of matrix A are the vectors

$$\mathbf{a}_1 = \begin{bmatrix} 2 \\ -2 \\ -2 \end{bmatrix}, \quad \mathbf{a}_2 = \begin{bmatrix} 4 \\ 3 \\ 4 \end{bmatrix}, \quad and \quad \mathbf{a}_3 = \begin{bmatrix} -3 \\ 2 \\ 1 \end{bmatrix},$$

which were used in Example 2. According to the definition, the matrix-vector product is computed by forming a linear combination of the columns of matrix A, using the entries of vector $\mathbf{x}$ as the scalars. Note that the vector $\mathbf{x}$ has three entries, one for each column of matrix A. Otherwise, the multiplication would not be possible. Thus,

$$A\mathbf{x} = \begin{bmatrix} 2 & 4 & -3 \\ -2 & 3 & 2 \\ -2 & 4 & 1 \end{bmatrix} \begin{bmatrix} 1 \\ 0 \\ -2 \end{bmatrix}$$

$$= 1 \begin{bmatrix} 2 \\ -2 \\ -2 \end{bmatrix} + 0 \begin{bmatrix} 4 \\ 3 \\ 4 \end{bmatrix} - 2 \begin{bmatrix} -3 \\ 2 \\ 1 \end{bmatrix}$$

$$= \begin{bmatrix} 8 \\ -6 \\ -4 \end{bmatrix}$$

To perform this same calculation in MATLAB, we first enter the matrix A. Since we have already entered the column vectors in Example 2, it is easier to create A with the command

```
>> A = [a1,a2,a3]
A =
     2     4    -3
    -2     3     2
    -2     4     1
```

Next we enter the vector $\mathbf{x} = [1, 0, -2]^T$ into MATLAB and perform the multiplication with the commands

```
>> x = [1;0;-2]; b = A*x
b =
     8
    -6
    -4
```

Notice that this command is simply $*$ and not $.*$. The operation $*$ is called *matrix multiplication*, and it is quite different from $.*$, which is the operation of array multiplication that we studied in Chapter 2.

It is important to notice that the same solution is found in Examples 2 and 3. Realizing that matrix-vector multiplication amounts to taking a linear combination of the columns of the matrix, with the entries of the vector for the scalars, is the key to recognizing where matrix multiplication can be used to solve problems.

Let's use these ideas to solve a system of equations.

Example 4. *Solve the following system of equations for x_1, x_2, and x_3.*

$$2x_1 + 4x_2 - 3x_3 = 8$$
$$-2x_1 + 3x_2 + 2x_3 = -6 \tag{11.7}$$
$$-2x_1 + 4x_2 + x_3 = -4$$

The three equalities in (11.7) can be written as the single vector equality

$$\begin{bmatrix} 2x_1 + 4x_2 - 3x_3 \\ -2x_1 + 3x_2 + 2x_3 \\ -2x_1 + 4x_2 + x_3 \end{bmatrix} = \begin{bmatrix} 8 \\ -6 \\ -4 \end{bmatrix}. \tag{11.8}$$

Standard vector algebra allows us to write the left-hand side of equation (11.8) as the linear combination

$$x_1 \begin{bmatrix} 2 \\ -2 \\ -2 \end{bmatrix} + x_2 \begin{bmatrix} 4 \\ 3 \\ 4 \end{bmatrix} + x_3 \begin{bmatrix} -3 \\ 2 \\ 1 \end{bmatrix} = \begin{bmatrix} 8 \\ -6 \\ -4 \end{bmatrix}. \tag{11.9}$$

Next, we notice that the linear combination on the left-hand side of equation (11.9) can be written as a matrix-vector product, giving us

$$\begin{bmatrix} 2 & 4 & -3 \\ -2 & 3 & 2 \\ -2 & 4 & 1 \end{bmatrix} \begin{bmatrix} x_1 \\ x_2 \\ x_3 \end{bmatrix} = \begin{bmatrix} 8 \\ -6 \\ -4 \end{bmatrix}. \tag{11.10}$$

Equation (11.10) is in the form $A\mathbf{x} = \mathbf{b}$, where

$$A = \begin{bmatrix} 2 & 4 & -3 \\ -2 & 3 & 2 \\ -2 & 4 & 1 \end{bmatrix} \quad \text{and} \quad \mathbf{b} = \begin{bmatrix} 8 \\ -6 \\ -4 \end{bmatrix}.$$

The matrix A is the same matrix found in Example 3, and it is the coefficient matrix for the system in (11.7). It is formed by recording the coefficients of x_1, x_2, and x_3 from the system of equations in (11.7). In this case we can solve the equation $A\mathbf{x} = \mathbf{b}$ by dividing both sides of the equation *on the left* by the matrix A, leading to $A \backslash A\mathbf{x} = A \backslash \mathbf{b}$, or $\mathbf{x} = A \backslash \mathbf{b}$. This is easily accomplished in MATLAB. The matrix A and the vector $\mathbf{b}$ were formed in the previous examples, so all we need to do is enter

```
>> x = A\b
x =
         1
         0
        -2
```

Once again we have demonstrated the ease with which MATLAB can solve mathematical problems. Indeed the method of the previous paragraph works in great generality. **However, it does not always work!** In order for it to work, it is necessary that the coefficient matrix be *nonsingular* (the term nonsingular will be defined later in this chapter). Unfortunately, many of the applications of linear systems to ordinary differential equations have coefficient matrices which are singular. Furthermore, since matrix multiplication is a rather complicated procedure, one can expect that matrix division is at least equally complicated. The above naive approach is hiding a lot of mathematics as we will soon see.

Matrix Multiplication

Before continuing with our search for a method of solving systems of equations, let's extend the idea of matrix-vector product to the product of two matrices. The operation of multiplying two matrices is called *matrix multiplication*. It might seem complicated, but it is easily defined using matrix-vector multiplication defined earlier, and demonstrated in Example 3.

Suppose that we have two matrices A and B and that the number of columns of A is equal to the number of rows of B. Let $\mathbf{b}_1, \mathbf{b}_2, \ldots, \mathbf{b}_p$ represent the columns of matrix B. We will denote this by the equation

$$B = \left[\mathbf{b}_1, \mathbf{b}_2, \ldots, \mathbf{b}_p\right].$$

Then the *matrix product* AB is defined by

$$AB = A\left[\mathbf{b}_1, \mathbf{b}_2, \ldots, \mathbf{b}_p\right],$$
$$= \left[A\mathbf{b}_1, A\mathbf{b}_2, \ldots, A\mathbf{b}_p\right].$$

Therefore, the matrix product AB is formed by multiplying each column of the matrix B by the matrix A.

Example 5. *Find the matrix product AB, where*

$$A = \begin{bmatrix} 1 & -1 & 1 \\ 0 & 1 & -2 \\ 1 & 0 & 1 \end{bmatrix} \quad \text{and} \quad B = \begin{bmatrix} 2 & 1 \\ 2 & -2 \\ 2 & 0 \end{bmatrix}.$$

The first column of the matrix B is

$$\mathbf{b}_1 = \begin{bmatrix} 2 \\ 2 \\ 2 \end{bmatrix}.$$

Therefore, the first column of the matrix product AB is $A\mathbf{b}_1$. This multiplication would not be possible if it were not for the fact that the number of rows of B equals the number of columns of A. We get

$$A\mathbf{b}_1 = \begin{bmatrix} 1 & -1 & 1 \\ 0 & 1 & -2 \\ 1 & 0 & 1 \end{bmatrix} \begin{bmatrix} 2 \\ 2 \\ 2 \end{bmatrix}$$
$$= 2\begin{bmatrix} 1 \\ 0 \\ 1 \end{bmatrix} + 2\begin{bmatrix} -1 \\ 1 \\ 0 \end{bmatrix} + 2\begin{bmatrix} 1 \\ -2 \\ 1 \end{bmatrix}$$
$$= \begin{bmatrix} 2 \\ -2 \\ 4 \end{bmatrix}$$

158

The second column vector in B is $\mathbf{b}_2 = [1, -2, 0]^T$. In a similar manner we compute that

$$A\mathbf{b}_2 = \begin{bmatrix} 1 & -1 & 1 \\ 0 & 1 & -2 \\ 1 & 0 & 1 \end{bmatrix} \begin{bmatrix} 1 \\ -2 \\ 0 \end{bmatrix} = 1 \begin{bmatrix} 1 \\ 0 \\ 1 \end{bmatrix} - 2 \begin{bmatrix} -1 \\ 1 \\ 0 \end{bmatrix} + 0 \begin{bmatrix} 1 \\ -2 \\ 1 \end{bmatrix} = \begin{bmatrix} 3 \\ -2 \\ 1 \end{bmatrix}.$$

Consequently,

$$\begin{aligned} AB &= A\,[\mathbf{b}_1, \mathbf{b}_2], \\ &= [A\mathbf{b}_1, A\mathbf{b}_2], \\ &= \begin{bmatrix} 2 & 3 \\ -2 & -2 \\ 4 & 1 \end{bmatrix}. \end{aligned}$$

MATLAB will easily duplicate this effort and relieve us of the drudgery of hand calculations.

```
>> A = [1 -1 1; 0 1 -2; 1 0 1]; B=[2 1;2 -2;2 0];
>> C = A*B
C =
        2      3
       -2     -2
        4      1
```

Remark: Note that it is possible to define AB only if the number of rows in B equals the number of columns in A. More precisely, If A is an $m \times n$ matrix and B is a $n \times p$ matrix, then the matrix multiplication is possible, and the product AB is a $m \times p$ matrix. We say that matrix multiplication is possible if "the inner dimensions match," and the "outer dimensions give the dimension of the product." In Example 5, A is 3×3 and B is 3×2. Since the inner dimensions are both 3, it is possible to multiply these matrices. The outer dimensions of matrix A and B predict that the matrix product will have 3 rows and 2 columns, which is precisely what we saw in Example 5.

Example 6. *If*

$$A = \begin{bmatrix} 1 & 2 \\ 3 & 4 \end{bmatrix} \quad \text{and} \quad B = \begin{bmatrix} 1 & 0 \\ -1 & 2 \\ 4 & -3 \end{bmatrix},$$

use MATLAB to find the matrix products AB and BA.

Matrix A is 2×2 and matrix B is 3×2. The inner dimensions do not match, so it is not possible to form the matrix product AB. We should expect MATLAB to complain, and it does:

```
>> A=[1,2;3,4];B=[1,0;-1,2;4,-3];
>> A*B
??? Error using ==> *
Inner matrix dimensions must agree.
```

However, matrix B is 3×2 and matrix A is 2×2. This time the inner dimensions match, so it is possible to form the product BA. Further, the outer dimensions of B and A predict that the matrix product BA

will be a 3×2 matrix.

```
>> B*A
ans =
        1     2
        5     6
       -5    -4
```

Example 7. *If*

$$A = \begin{bmatrix} 1 & 2 \\ 3 & 4 \end{bmatrix} \quad and \quad B = \begin{bmatrix} 1 & -1 \\ -1 & 1 \end{bmatrix},$$

use MATLAB to find AB and BA and comment on the result.

Enter the matrices and perform the calculations.

```
>> A = [1,2;3,4]; B = [1,-1;-1,1];
>> A*B, B*A
ans =
       -1     1
       -1     1
ans =
       -2    -2
        2     2
```

Since $AB \neq BA$, this example clearly indicates that matrix multiplication is *not commutative*. If you switch the order of multiplication, you cannot expect that the answer will be the same. That is why we were careful to divide both sides of the equation $Ax = b$ on the left to derive the answer $\mathbf{x} = A\backslash\mathbf{b}$ in Example 4.

We will explore further properties of matrices in the exercises.

Solving Systems of Equations and Row Operations

In Example 1, we transformed the original system

$$\begin{aligned} u + 2v &= 5, \\ 4u - v &= 2. \end{aligned} \tag{11.11}$$

into the system

$$\begin{aligned} u + 2v &= 5, \\ -9v &= -18, \end{aligned} \tag{11.12}$$

by solving and substituting. The easier system in (11.12) was then solved by backsolving. It is difficult to implement the process of solving and substituting for larger systems. The key to developing a more systematic approach is to notice that we can get the system in (11.12) by subtracting 4 times the first equation in (11.11) from the second equation in (11.11). This operation of adding (or subtracting) a multiple of one equation to (or from) another is equivalent to solving and substituting, but is more direct

160

and easier to implement. It is easily seen that such an operation on an arbitrary system leads to a new system which is equivalent to the first in the sense that any solution of one system is also a solution of the other.

Our next step in devising a general solution method is to use matrix notation. For example, we have seen that the system in equation (11.11) can be written as the matrix equation

$$\begin{bmatrix} 1 & 2 \\ 4 & -1 \end{bmatrix} \begin{bmatrix} u \\ v \end{bmatrix} = \begin{bmatrix} 5 \\ 2 \end{bmatrix} \qquad (11.13)$$

The matrix equation (11.13) has the form $A\mathbf{x} = \mathbf{b}$, where

$$A = \begin{bmatrix} 1 & 2 \\ 4 & -1 \end{bmatrix}, \quad \mathbf{x} = \begin{bmatrix} u \\ v \end{bmatrix}, \quad \text{and} \quad \mathbf{b} = \begin{bmatrix} 5 \\ 2 \end{bmatrix}.$$

All of the information about the system is contained in the coefficient matrix A and the right-hand side vector $\mathbf{b}$. From these two we form the *augmented matrix* M by adding the column vector $\mathbf{b}$ as a third column to the matrix A; i.e.,

$$M = [A, \mathbf{b}] = \begin{bmatrix} 1 & 2 & 5 \\ 4 & -1 & 2 \end{bmatrix}.$$

The augmented matrix contains all of the information about the system in a compact form.

Notice that each row in M corresponds to one of the equations in (11.11). For example, the first row of M, [1 2 5], represents the equation $u + 2v = 5$. Furthermore, if we add -4 times the first row of M to the second row, we get

$$\begin{bmatrix} 1 & 2 & 5 \\ 0 & -9 & -18 \end{bmatrix},$$

which is the augmented matrix corresponding to the system in (11.12). Thus, we see that the method of solving and substitution in the system of equations becomes the operation of adding a multiple of one row of the augmented matrix to another row.

This is our first example of a *row operation* on a matrix. There are three in all.

Theorem 1. *Each of the following elementary row operations, when applied to the augmented matrix of a system of equations, will transform the augmented matrix into the augmented matrix of another system of equations which has exactly the same solutions as the original system.*

R1. *Add a multiple of one row to another row.*
R2. *Interchange two rows.*
R3. *Multiply a row by a nonzero constant.*

If the matrix is the augmented matrix associated to a system of equations, then operation R2 corresponds to interchanging two of the equations, and R3 corresponds to multiplying an equation by a nonzero constant.

Our general strategy for solving a system of linear equations is to replace the system by its augmented matrix, and then perform row operations to eliminate coefficients[2] until the system associated with the transformed matrix is easy to solve by the method of backsolving. Of course, each of the row operations requires a significant amount of arithmetic, but we will let MATLAB do that arithmetic for us. We will discuss how to do this in MATLAB after addressing some preliminaries.

---

[2] In this case, "eliminate" means "make equal to zero."

161

Matrix Indexing in MATLAB and Row Operations

One of the key features of MATLAB is the ease with which the elements of a matrix can be accessed and manipulated. If M is a matrix in MATLAB, then the entry in the second row and fourth column is denoted by M(2,4). The basic idea is that the first index refers to the row, and the second to the column. Try the following example.

```
>> M = [1 3 5 6;3 5 0 -3;-4 0 9 3]
M =
        1       3       5       6
        3       5       0      -3
       -4       0       9       3

>> M(2,4)
ans =
       -3
```

Try M(1,3). How would you access the entry in row three, column four, of matrix M?

The entries in a matrix can be easily changed. For example, if we want to change M(3,4) to -15, we enter

```
>> M(3,4) = -15
M =
        1       3       5       6
        3       5       0      -3
       -4       0       9     -15
```

and the job is done.

We can refer to the second row in the matrix M by M(2,:). The 2 refers to the second row, while the colon means we want all columns. In other words, we want the entire second row.

```
>> M(2,:)
ans =
        3       5       0      -3
```

Similarly, the command >> M(:,2) displays all rows and the second column, or the entire second column of M. Try it!

We can also easily refer to submatrices. For example, if we want to refer to the matrix consisting of the first and third rows of M, we enter

```
>> M([1 3],:)
ans =
        1       3       5       6
       -4       0       9     -15
```

and we get what we want. Of course, we can do the same with columns.

```
>> M(:,[2,4])
ans =
        3       6
        5      -3
        0     -15
```

What do you think M([1,3],[4,1]) refers to? Try it and see.

Now we are ready to explain how to do row operations. We will start with the matrix M we ended up with above, i.e.

```
>> M
M =
        1       3       5       6
        3       5       0      -3
       -4       0       9     -15
```

We will illustrate the operation R1 by adding 4 times the first row to the third row. The notation for the third row is M(3,:), so we will replace this by M(3,:) + 4*M(1,:). We will let MATLAB do all of the tedious arithmetic.

```
>> M(3,:) = M(3,:) + 4*M(1,:)
M =
        1       3       5       6
        3       5       0      -3
        0      12      29       9
```

To illustrate operation R2 we will exchange rows 2 and 3. In this case we want M([2,3],:) to be equal to what M([3,2],:) is currently, so we execute the following command

```
>> M([2,3],:) = M([3,2],:)
M =
        1       3       5       6
        0      12      29       9
        3       5       0      -3
```

Finally, we illustrate R3 by multiplying row 2 by -5:

```
>> M(2,:) = -5*M(2,:)
M =
        1       3       5       6
        0     -60    -145     -45
        3       5       0      -3
```

We can divide a row by a number just as easily:

```
>> M(3,:) = M(3,:)/3
M =
     1.0000    3.0000    5.0000    6.0000
          0  -60.0000 -145.0000  -45.0000
     1.0000    1.6667         0   -1.0000
```

The Rational Format

Frequently problems arise in which the numbers we are using are all rational numbers. The floating point arithmetic that MATLAB uses does not lend itself to maintaining the rational form of numbers. We can sometimes recognize a rational number from the decimal expansions MATLAB gives us, but frequently, we can't. For example, in M as it appears above, we know that M(3,2) = 5/3, but we will not always be so fortunate. MATLAB provides a solution with its rational format. If we enter format rat at the MATLAB prompt, from that point on MATLAB will display numbers as the ratios of integers. For example,

```
>> format rat
>> pi
ans =
     355/113
```

We know, and MATLAB knows, that $\pi \neq 355/113$. MATLAB is only displaying an approximation to the answer. But this is no different than what MATLAB always does. The only difference is that now MATLAB is using a rational number to approximate π instead of a decimal expansion. It computes and displays the rational number that is closest to the given number within a prescribed tolerance. The advantage is that if we know the answer is a rational number, then we know that the rational representation shown us by MATLAB is probably 100% correct.

In the rational format, our current matrix M is

```
>> M
M =
     1        3        5        6
     0      -60     -145      -45
     1      5/3        0       -1
```

Thus we get the precise, rational version of M.

We should warn you about one possibly confusing aspect of the rational format. Sometimes asterisks (*) appear in the output. This happens when MATLAB is trying to give a rational approximation to a very small number. A very large denominator is required, and the resulting expression will not fit within the rational format. When we know that the matrix has entries that are rational numbers with small denominators, the small number represented by an asterisk is due to round off error. The entry should actually be 0, and it can usually be replaced by 0, but caution is called for in doing so.

Elimination and Backsolving

The first part of our strategy for solving systems is to use row operations to put the system into a form which can be easily solved by backsolving. The goal is a matrix in *row echelon form*. This is a matrix that looks like

$$\begin{bmatrix} P & \# & \# & \# & \# & \# & \# & \# & \# & \# \\ 0 & P & \# & \# & \# & \# & \# & \# & \# & \# \\ 0 & 0 & 0 & 0 & P & \# & \# & \# & \# & \# \\ 0 & 0 & 0 & 0 & 0 & P & \# & \# & \# & \# \\ 0 & 0 & 0 & 0 & 0 & 0 & 0 & 0 & P & \# \\ 0 & 0 & 0 & 0 & 0 & 0 & 0 & 0 & 0 & 0 \\ 0 & 0 & 0 & 0 & 0 & 0 & 0 & 0 & 0 & 0 \end{bmatrix}, \qquad (11.14)$$

where the Ps stand for nonzero numbers and the #s stand for arbitrary numbers. We will define the *pivot* of a row in a matrix to be the first nonzero number in that row. Thus the Ps in (11.14) are pivots. The distinguishing feature of a matrix in row echelon form is that, in any row that has a pivot, the pivot lies strictly to the right of the pivots in preceding rows.

After working some examples, and some thought, you should be able to convince yourself that any matrix can be reduced to row echelon form using row operations. In fact, we can go further. The pivots can all be reduced to 1 by dividing each row of the matrix in (11.14) by its pivot. In addition, all of the nonzero entries above a pivot can be reduced to 0 by adding (or subtracting) a multiple of the pivot row. In this way we can reduce the matrix to the form

$$\begin{bmatrix} 1 & 0 & \# & \# & 0 & 0 & \# & \# & 0 & \# \\ 0 & 1 & \# & \# & 0 & 0 & \# & \# & 0 & \# \\ 0 & 0 & 0 & 0 & 1 & 0 & \# & \# & 0 & \# \\ 0 & 0 & 0 & 0 & 0 & 1 & \# & \# & 0 & \# \\ 0 & 0 & 0 & 0 & 0 & 0 & 0 & 0 & 1 & \# \\ 0 & 0 & 0 & 0 & 0 & 0 & 0 & 0 & 0 & 0 \\ 0 & 0 & 0 & 0 & 0 & 0 & 0 & 0 & 0 & 0 \end{bmatrix}, \qquad (11.15)$$

where the #'s stand for arbitrary numbers. A matrix in the form of (11.15) is said to be in *reduced row echelon* form.

Our entire strategy for solving systems of linear equations amounts to taking the augmented matrix of the system and transforming it to row echelon form using elementary row operations. This method is called *Gaussian elimination*. You will learn that a system in row echelon form can be easily solved by backsolving.

Example 8. *Find all solutions to the system*

$$\begin{aligned} x_2 - x_3 + 6x_4 &= 0, \\ 3x_1 - 4x_2 + 5x_3 - x_4 &= 5, \\ 5x_1 - x_3 + 4x_4 &= 4, \end{aligned} \qquad (11.16)$$

for x_1, x_2, x_3, and x_4.

We first write the system in (11.16) in the matrix form $A\mathbf{x} = \mathbf{b}$, where

$$A = \begin{bmatrix} 0 & 1 & -1 & 6 \\ 3 & -4 & 5 & -1 \\ 5 & 0 & -1 & 4 \end{bmatrix}, \quad \mathbf{x} = \begin{bmatrix} x_1 \\ x_2 \\ x_3 \\ x_4 \end{bmatrix}, \quad \text{and} \quad \mathbf{b} = \begin{bmatrix} 0 \\ 5 \\ 4 \end{bmatrix}.$$

In MATLAB, we first enter A and $\mathbf{b}$ and form the augmented matrix:

```
>> A = [0 1 -1 6;3 -4 5 -1;5 0 -1 4]; b=[0 5 4]'; M = [A,b]
M =
      0            1           -1            6            0
      3           -4            5           -1            5
      5            0           -1            4            4
```

Reduction to row echelon form. To start this process we want to get a pivot into the upper left-hand corner. To do so, we use operation $R2$, and interchange the first and second rows:

```
>> M([1,2],:) = M([2,1],:)
M =
      3           -4            5           -1            5
      0            1           -1            6            0
      5            0           -1            4            4
```

Now that the pivot in the first row is $M(1,1) = 3$, we want to make all of the entries below it equal to 0 in order to move the pivots in the lower rows to the right. Since $M(2,1) = 0$, we need do nothing to the second row. For the third row we subtract 5/3 times the first row to make $M(3,1) = 0$:

```
>> M(3,:) = M(3,:) - (5/3)*M(1,:)
M =
      3           -4            5           -1            5
      0            1           -1            6            0
      0          20/3        -28/3         17/3        -13/3
```

The pivot in the second row is $M(2,2) = 1$ and that in the third row is $M(3,2) = 20/3$. We want to make $M(3,2) = 0$ in order to move the pivot in the third row to the right. We can do so by subtracting 20/3 times the second row from the third:

```
>> M(3,:) = M(3,:) - (20/3)*M(2,:)
M =
      3           -4            5           -1            5
      0            1           -1            6            0
      0            0         -8/3       -103/3        -13/3
```

The matrix is now in row echelon form, but not in reduced row echelon form. It is the augmented matrix for the system

$$3x_1 - 4x_2 + 5x_3 - x_4 = 5$$
$$x_2 - x_3 + 6x_4 = 0 \tag{11.17}$$
$$-(8/3)x_3 - (103/3)x_4 = -13/3$$

166

The system in (11.17) has the same solutions as the system in (11.16), but this one can be easily solved by backsolving.

Backsolving. We start by solving the last equation in (11.17). This involves only the two variables x_3 and x_4. We can assign an arbitrary value to either one and solve for the other. We need a systematic way of deciding which variable is assigned the arbitrary value.

We will call a column in a matrix in row echelon form which contains a pivot a *pivot column*. The corresponding variables are called *pivot variables*. In our example, columns 1, 2, and 3 are pivot columns, and x_1, x_2, and x_3 are the pivot variables. A column which does not contain a pivot is called a *free column*. The variables corresponding to free columns are called *free variables*. Thus column 4 is a free column and x_4 is a free variable. Notice that we can assign an arbitrary value to the free variable x_4 and still solve (11.17) for the pivot variables. Consequently, we set $x_4 = t$. Then solving the third equation in (11.17) for x_3 we get

$$x_3 = (13 - 103x_4)/8 = (13 - 103t)/8.$$

Now we can solve the second equation for x_2:

$$x_2 = x_3 - 6x_4 = (13 - 108t)/8 - 6t = (13 - 151t)/8.$$

Finally, we can solve the first equation for x_1:

$$
\begin{aligned}
x_1 &= (5 + 4x_2 - 5x_3 + x_4)\,/3 \\
&= (5 + 4(13 - 151t)/8 - 5(13 - 103t)/8 + t)\,/3 \\
&= (9 - 27t)/8.
\end{aligned}
$$

We can sum up our results in vector form. The solutions to (11.17), and therefore to (11.16), are all vectors of the form

$$
\mathbf{x} = \begin{bmatrix} x_1 \\ x_2 \\ x_3 \\ x_4 \end{bmatrix} = \begin{bmatrix} (9 - 27t)/8 \\ (13 - 151t)/8 \\ (13 - 103t)/8 \\ t \end{bmatrix} = \begin{bmatrix} 9/8 \\ 13/8 \\ 13/8 \\ 0 \end{bmatrix} + t \begin{bmatrix} -27/8 \\ -151/8 \\ -103/8 \\ 1 \end{bmatrix}. \tag{11.18}
$$

This solution is called the *general* solution, since all possible solutions of system (11.16) can be found by substituting arbitrary numbers for the parameter t. For example, $t = 0$ yields the solution $x_1 = 9/8$, $x_2 = 13/8$, $x_3 = 13/8$, and $x_4 = 0$; $t = 1$ yields the solution $x_1 = -9/4$, $x_2 = -69/4$, $x_3 = -45/4$, and $x_4 = 1$; etc.

Reduction to reduced row echelon form. For the sake of comparison, let's reduce the augmented matrix M to reduced row echelon form. We will learn an important insight along the way. We need to make the pivots equal to 1, and this is accomplished by dividing each row by its pivot:

```
>> M(1,:) = M(1,:)/3; M(3,:) = (-3/8)*M(3,:);
```

167

In addition we have to make the entries above the pivots equal to 0. We can do this by adding multiples of a row to those above it. The operations to make $M(1,2) = 0$ and $M(1,3) = 0$ are

```
>> M(1,:) = M(1,:) +(4/3)*M(2,:);
>> M(1,:) = M(1,:) -(1/3)*M(3,:)
M =
         1              0              *           27/8           9/8
         0              1             -1              6             0
         0              0              1          103/8          13/8
```

Here we have a surprise. The asterisk (*) indicates that the entry, $M(1,3)$, is a rational number whose rational form is too large to display on the screen. Remember that $M(1,3)$ should be equal to 0. To see what it really is we execute

```
>> M(1,3)
ans =
       1/6004799503160662
```

Thus $M(1,3)$ is a very small number. MATLAB is making small roundoff errors all the time, and this is the result. We have been a little careless. When we do row operations, we will get better precision if we refer to the matrix elements by name instead of using the number that MATLAB is using to display them on the screen. For example, if we execute

```
>> M(1,:) = M(1,:) -M(1,3)*M(3,:)
M =
         1              0              0           27/8           9/8
         0              1             -1              6             0
         0              0              1          103/8          13/8
```

we are back on track. Notice that we referred directly to $M(1,3)$ in this command. This is the preferred method for doing row operations. We will use this procedure in the future, and we recommend that you do too. For example, the command to make $M(2,3)$ equal to 0 is

```
>> M(2,:) = M(2,:) - M(2,3)*M(3,:)
M =
         1              0              0           27/8           9/8
         0              1              0          151/8          13/8
         0              0              1          103/8          13/8
```

which gives us the reduced row echelon form for the augmented matrix.

This matrix is the augmented matrix for the system

$$
\begin{aligned}
x_1 + (27/8)x_4 &= 9/8, \\
x_2 + (151/8)x_4 &= 13/8, \\
x_3 + (103/8)x_4 &= 13/8.
\end{aligned}
\tag{11.19}
$$

Of course, the system in (11.19) has the same solutions as the systems in (11.16) and (11.17). Solving to find the solutions given in (11.18) is extremely easy. However, we had to do five additional row operations to get to reduced row echelon form and the easier system in (11.19). This raises a question. Starting at the row echelon form we have two ways to proceed. We can either solve the system given by the row echelon form, or we can perform the additional row operations to get to reduced row echelon form and the easier system. Both give the correct answer, but which is the easiest method? This is a hard question to answer, but most people find it preferable to stop at row echelon form, and not do the additional work of reducing the augmented matrix to reduced row echelon form. Try it both ways, and choose the method which is easiest for you.

The method we used on Example 8 works for any system. Let's list the steps in the process.

Summary: Method of solution for a system of linear equations.

1. Write the system in the matrix form $A\mathbf{x} = \mathbf{b}$ and form the augmented matrix $M = [A, \mathbf{b}]$.
2. Use row operations to reduce M to row echelon form.
3. Write down the simplified system.
4. Solve the simplified system by backsolving:
    - Assign arbitrary values to the free variables.
    - Backsolve for the pivot variables.

The MATLAB command rref. After completing Example 8, you must be wondering why the computer and MATLAB can't do step 2 in the solution process, the reduction of a matrix to row echelon form, or even to reduced row echelon form, with a single command. In fact it can. If we start with the original augmented matrix

```
>> M = [A,b];
```

then we can proceed directly to the reduced row echelon form by using the MATLAB command `rref`.

```
>> MR = rref(M)
MR =
        1        0        0      27/8      9/8
        0        1        0     151/8     13/8
        0        0        1     103/8     13/8
```

This eliminates all of the intermediate steps, and we can proceed to the third step and write down the equivalent system of equations as we did in (11.19).

Example 9. *Find all solutions to the system*

$$2x_2 + 2x_3 + 3x_4 = -4,$$
$$-2x_1 + 4x_2 + 2x_3 - x_4 = -6,$$
$$3x_1 - 4x_2 - x_3 + 2x_4 = 8.$$

The augmented matrix is

```
>> M = [0 2 2 3 -4; -2 4 2 -1 -6; 3 -4 -1 2 8]
M =
         0            2            2            3           -4
        -2            4            2           -1           -6
         3           -4           -1            2            8
```

Place the augmented matrix in reduced row echelon form:

```
>> rref(M)
ans =
         1            0            1            0         16/5
         0            1            1            0         -1/5
         0            0            0            1         -6/5
```

Notice that columns 1, 2, and 4 are pivot columns, so x_1, x_2, and x_4 are pivot variables and x_3 is a free variable. The equivalent system is

$$x_1 + x_3 = 16/5,$$
$$x_2 + x_3 = -1/5,$$
$$x_4 = -6/5.$$

Solve each of the above equations for a pivot variable, in terms of the free variable x_3.

$$x_1 = 16/5 - x_3$$
$$x_2 = -1/5 - x_3$$
$$x_4 = -6/5$$

Set $x_3 = t$, where t is an arbitrary number, and we see that the solutions of the system are all vectors of the form

$$\mathbf{x} = \begin{bmatrix} x_1 \\ x_2 \\ x_3 \\ x_4 \end{bmatrix} = \begin{bmatrix} 16/5 - t \\ -1/5 - t \\ t \\ -6/5 \end{bmatrix} = \begin{bmatrix} 16/5 \\ -1/5 \\ 0 \\ -6/5 \end{bmatrix} + t \begin{bmatrix} -1 \\ -1 \\ 1 \\ 0 \end{bmatrix},$$

where t is arbitrary.

Determined Systems of Equations

A system is *determined* if it has the same number of equations and unknowns. It is *underdetermined* if it has fewer equations than unknowns, and *overdetermined* if it has more equations than unknowns. Some of the systems we have been solving up to now have been underdetermined. In this section we examine determined systems.

There are three possibilities for the set of all solutions to a system of linear equations:

- there are no solutions,
- there are infinitely many solutions,
- there is a unique solution.

170

The next three examples of determined systems illustrate these possibilities.

Example 10. *Find all solutions to the system*

$$4x_1 - 4x_2 - 8x_3 = 27,$$
$$2x_2 + 2x_3 = -6, \tag{11.20}$$
$$x_1 - 2x_2 - 3x_3 = 10.$$

The system in (11.20) is determined, since it has three unknowns (x_1, x_2, and x_3) and three equations. Set up the augmented matrix

```
>> M = [4 -4 -8 27;0 2 2 -6;1 -2 -3 10]
M =
     4          -4          -8          27
     0           2           2          -6
     1          -2          -3          10
```

and find its reduced row echelon form.

```
>> rref(M)
ans =
     1           0          -1           0
     0           1           1           0
     0           0           0           1
```

The last row of the reduced row echelon form represents the equation $0x_1 + 0x_2 + 0x_3 = 1$, which clearly has no solutions. If there is no solution which satisfies this last equation, then there certainly cannot be a solution which satisfies all three equations represented by the matrix. Consequently, the system (11.20) has no solutions. A system that has solutions is called a *consistent* system. Since this system has no solutions it is *inconsistent*.

Example 11. *Find all solutions to the system*

$$4x_1 - 4x_2 - 8x_3 = 4,$$
$$2x_2 + 2x_3 = 2, \tag{11.21}$$
$$x_1 - 2x_2 - 3x_3 = 0.$$

The system in (11.21) is determined, since it has three unknowns (x_1, x_2, and x_3) and three equations. Set up the augmented matrix

```
>> M = [4 -4 -8 4;0 2 2 2;1 -2 -3 0]
M =
     4     -4     -8      4
     0      2      2      2
     1     -2     -3      0
```

and find its reduced row echelon form.

```
>> rref(M)
ans =
     1     0    -1     2
     0     1     1     1
     0     0     0     0
```

Columns 1 and 2 are pivot columns, while column 3 is free. Thus x_3 is a free variable. The equivalent system is

$$x_1 - x_3 = 2,$$
$$x_2 + x_3 = 1. \tag{11.22}$$

Note that the last row of the matrix, $[0\ 0\ 0\ 0]$, represents the equation $0x_1 + 0x_2 + 0x_3 = 0$. Since any combination of x_1, x_2, and x_3 is a solution of this equation, we need only turn our attention to finding solutions of system (11.22). Solve each equation of system (11.22) for a pivot variable in terms of the free variable x_3.

$$x_1 = 2 + x_3$$
$$x_2 = 1 - x_3$$

If we set $x_3 = t$, where t is an arbitrary number, and place the solution in vector form, then

$$\mathbf{x} = \begin{bmatrix} x_1 \\ x_2 \\ x_3 \end{bmatrix} = \begin{bmatrix} 2+t \\ 1-t \\ t \end{bmatrix} = \begin{bmatrix} 2 \\ 1 \\ 0 \end{bmatrix} + t \begin{bmatrix} 1 \\ -1 \\ 1 \end{bmatrix},$$

where t is an arbitrary number. Thus system (11.21) has infinitely many solutions.

Example 12. *Find all solutions to the system*

$$3x_1 - 4x_2 - 8x_3 = 27,$$
$$2x_2 + 2x_3 = -6, \tag{11.23}$$
$$x_1 - 2x_2 - 3x_3 = 10.$$

Enter the augmented matrix

```
>> M = [3 -4 -8 27;0 2 2 -6;1 -2 -3 10]
M =
     3    -4    -8    27
     0     2     2    -6
     1    -2    -3    10
```

and place it in reduced row echelon form.

```
>> rref(M)
ans =
     1     0     0     1
     0     1     0     0
     0     0     1    -3
```

172

We see that the only solution of system (11.23) is $x_1 = 1$, $x_2 = 0$, and $x_3 = -3$.

These examples show that the situation with determined systems is not so simple. Sometimes there are solutions, sometimes not. Sometimes solutions are unique, sometimes not. We need a way to tell which phenomena occur in specific cases. The answer is provided by the *determinant.*

The Determinant and Systems

We will not study the determinant in any detail. Let's just recall some facts. The determinant is defined for any square matrix, i.e. any matrix which has the same number of rows and columns. For two by two and three by three matrices we have the formulas

$$\det \begin{bmatrix} a & b \\ c & d \end{bmatrix} = ad - bc$$

and

$$\det \begin{bmatrix} a & b & c \\ d & e & f \\ g & h & i \end{bmatrix} = aei - afh - bdi + cdh + bfg - ceg.$$

For larger matrices the formula for the determinant gets increasingly lengthy, and decreasingly useful for calculation. Once more MATLAB comes to our rescue, since it has a built-in procedure det for calculating the determinants of matrices of arbitrary size.

Consider the determinant of the coefficient matrix of systems (11.20) and (11.21) of Examples 10 and 11.

```
>> A = [4 -4 -8;0 2 2;1 -2 -3]
A =
     4    -4    -8
     0     2     2
     1    -2    -3
>> det(A)
ans =
     0
```

Note that the determinant is zero. On the other hand, for the coefficient matrix for system (11.23) of Example 12, we have

```
>> A = [3 -4 -8;0 2 2;1 -2 -3]
A =
     3    -4    -8
     0     2     2
     1    -2    -3
>> det(A)
ans =
     2
```

The fact that the determinant of the coefficient matrix of system (11.23) is nonzero and the solution is unique is no coincidence.

Theorem 2. *Let* **A** *be an* $n \times n$ *matrix.*

a) *If* det $A \neq 0$, *then for every column* n*-vector* **b** *there is a unique vector* **x** *such that* $A\mathbf{x} = \mathbf{b}$. *In particular, the only solution to the homogeneous equation* $A\mathbf{x} = \mathbf{0}$ *is the zero vector,* $\mathbf{x} = \mathbf{0}$.

b) *If* det$(A) = 0$, *then there are vectors* **b** *for which there are no vectors* **x** *with* $A\mathbf{x} = \mathbf{b}$. *If there is a solution to the equation* $A\mathbf{x} = \mathbf{b}$, *then there are infinitely many. In particular, the homogeneous equation* $A\mathbf{x} = \mathbf{0}$ *has nonzero solutions.*

If det $A = 0$, we will say that the matrix A is *singular*. As we saw in Examples 10 and 11, if the coefficient matrix is singular, then two possibilities exist. Either the system $A\mathbf{x} = \mathbf{b}$ has no solutions, or it has infinitely many solutions.

On the other hand, if det $A \neq 0$, we will say that A is *nonsingular*. If the matrix A is nonsingular, then the system $A\mathbf{x} = \mathbf{b}$ has a unique solution for every choice of **b**. MATLAB has a convenient way to find this solution. It is only necessary to enter x = A\b. Hence, system (11.23) can be solved as follows:

```
>> A = [3 -4 -8;0 2 2;1 -2 -3];b=[27;-6;10];
>> x = A\b
x =
        1
   1/1125899906842624
       -3
```

The very large denominator in the second term indicates that MATLAB is approximating 0. (Look at this answer in format long.) Thus, this solution agrees with the one found in Example 12. On the other hand, if we try the same technique on system (11.20) of Example 10, we get

```
>> A = [4 -4 -8;0 2 2;1 -2 -3];b=[27;-6;10];
>> x = A\b

Warning: Matrix is singular to working precision.
x =
      1/0
      1/0
      1/0
```

we see that MATLAB recognizes the fact that A is a singular matrix.

The backslash command A\b is deceptively simple. It looks as if we are dividing the equation $A\mathbf{x} = \mathbf{b}$ on the left by the matrix A to produce the solution x = A\b. In a way this is true, but in order to implement that division, MATLAB has to go through a series of computations involving row operations which is very similar to those discussed in this chapter.

If the matrix A is nonsingular, then it has an *inverse*. This is another matrix B such that $AB = BA = I$. Here I is the *identity matrix*, which has all ones along the diagonal, and zeros off the diagonal. In general it is time-consuming to calculate the inverse of a matrix, but again MATLAB can do this for us. We enter B = inv(A). For example, if A is the coefficient matrix of system (11.23), then

```
>> A = [3 -4 -8;0 2 2;1 -2 -3]
A =
        3              -4             -8
        0               2              2
        1              -2             -3
>> B = inv(A)
B =
       -1               2              4
        1            -1/2             -3
       -1               1              3
```

You can verify that $BA = AB = I$, with B*A and A*B. Try it! By the way, in MATLAB the command eye(n) is used to designate the $n \times n$ identity matrix. Try eye(3).

The Nullspace of a Matrix

The *nullspace* of a matrix A is the set of all vectors $\mathbf{x}$ such that $A\mathbf{x} = \mathbf{0}$. Because $\mathbf{x} = \mathbf{0}$ is always a solution of $A\mathbf{x} = \mathbf{0}$, the zero vector is always an element of the nullspace of matrix A. However, if A is nonsingular, then Theorem 2 guarantees that this solution is unique, and the nullspace of A consists of only the zero vector $\mathbf{0}$. On the other hand, if matrix A is singular, then the nullspace will contain lots of non-zero vectors.

Example 13. *Find the nullspace of matrix*

$$A = \begin{bmatrix} 12 & -8 & -4 & 30 \\ 3 & -2 & -1 & 6 \\ 18 & -12 & -6 & 42 \\ 6 & -4 & -2 & 15 \end{bmatrix}.$$

The nullspace of A consists of all solutions of $A\mathbf{x} = \mathbf{0}$, or

$$\begin{bmatrix} 12 & -8 & -4 & 30 \\ 3 & -2 & -1 & 6 \\ 18 & -12 & -6 & 42 \\ 6 & -4 & -2 & 15 \end{bmatrix} \begin{bmatrix} x_1 \\ x_2 \\ x_3 \\ x_4 \end{bmatrix} = \begin{bmatrix} 0 \\ 0 \\ 0 \\ 0 \end{bmatrix}. \tag{11.24}$$

Set up the augmented matrix with the commands

```
>> A = [12 -8 -4 30;3 -2 -1 6;18 -12 -6 42;6 -4 -2 15];
>> b = zeros(4,1);
>> M = [A,b];
```

and find its reduced row echelon form,

```
>> rref(M)
ans =
        1            -2/3            -1/3              0              0
        0               0               0              1              0
        0               0               0              0              0
        0               0               0              0              0
```

175

Note that the pivot columns are one and four, making x_1 and x_4 pivot variables, leaving x_2 and x_3 as free variables. The equivalent system is

$$x_1 - (2/3)x_2 - (1/3)x_3 = 0,$$
$$x_4 = 0. \tag{11.25}$$

Solve each equation of system (11.25) for a pivot variable, in terms of the free variables x_2 and x_3.

$$x_1 = (2/3)x_2 + (1/3)x_3$$
$$x_4 = 0$$

Set $x_2 = s$ and $x_3 = t$, where s and t are arbitrary numbers, then write the solution in vector form.

$$\mathbf{x} = \begin{bmatrix} x_1 \\ x_2 \\ x_3 \\ x_4 \end{bmatrix} = \begin{bmatrix} (2/3)s + (1/3)t \\ s \\ t \\ 0 \end{bmatrix} = \begin{bmatrix} (2/3)s \\ s \\ 0 \\ 0 \end{bmatrix} + \begin{bmatrix} (1/3)t \\ 0 \\ t \\ 0 \end{bmatrix} = s \begin{bmatrix} 2/3 \\ 1 \\ 0 \\ 0 \end{bmatrix} + t \begin{bmatrix} 1/3 \\ 0 \\ 1 \\ 0 \end{bmatrix}$$

Thus, every element in the nullspace of matrix A can be written as a linear combination of the vectors $\mathbf{v}_1 = [2/3, 1, 0, 0]^T$ and $\mathbf{v}_2 = [1/3, 0, 1, 0]^T$. The nullspace contains infinitely many vectors, but if we know $\mathbf{v}_1$ and $\mathbf{v}_2$ we know them all.

The results in this example are typical of what happens in general. The nullspace of a matrix will consist of all linear combinations of a small number of special vectors. If these vectors are linearly independent (see the next section), they are called a *basis* of the nullspace, and the nullspace is said to be *spanned* by the basis vectors. If we know a basis for a nullspace, then we know all of the vectors in the nullspace. For this reason we will almost always describe a nullspace by giving a basis.

Linear Dependence and Independence

A finite set of vectors is said to be *linearly independent* if the only linear combination which is equal to the zero vector is the one where all of the coefficients are equal to zero; i.e., the vectors $\mathbf{v}_1, \mathbf{v}_2, \cdots, \mathbf{v}_p$ are linearly independent if $c_1\mathbf{v}_1 + c_2\mathbf{v}_2 + \cdots + c_p\mathbf{v}_p = \mathbf{0}$ implies that $c_1 = c_2 = \cdots = c_p = 0$. A set of vectors which is not linearly independent is said to be *linearly dependent*.

Example 14. *Show that the vectors*

$$\mathbf{v}_1 = \begin{bmatrix} 1 \\ 1 \\ 1 \end{bmatrix}, \quad \mathbf{v}_2 = \begin{bmatrix} 0 \\ -1 \\ 2 \end{bmatrix}, \quad and \quad \mathbf{v}_3 = \begin{bmatrix} 2 \\ -1 \\ 8 \end{bmatrix}$$

are linearly dependent.

We need to find a non-trivial solution of

$$c_1 \begin{bmatrix} 1 \\ 1 \\ 1 \end{bmatrix} + c_2 \begin{bmatrix} 0 \\ -1 \\ 2 \end{bmatrix} + c_3 \begin{bmatrix} 2 \\ -1 \\ 8 \end{bmatrix} = \begin{bmatrix} 0 \\ 0 \\ 0 \end{bmatrix}. \tag{11.26}$$

176

The vector equation (11.26) is equivalent to the matrix equation

$$\begin{bmatrix} 1 & 0 & 2 \\ 1 & -1 & -1 \\ 1 & 2 & 8 \end{bmatrix} \begin{bmatrix} c_1 \\ c_2 \\ c_3 \end{bmatrix} = \begin{bmatrix} 0 \\ 0 \\ 0 \end{bmatrix}. \tag{11.27}$$

Set up the augmented matrix and reduce.

```
>> A = [1 0 2;1 -1 -1;1 2 8]; b = zeros(3,1);
>> M = [A,b];
>> rref(M)
ans =
       1         0         2         0
       0         1         3         0
       0         0         0         0
```

The solution is

$$c_1 = -2c_3,$$
$$c_2 = -3c_3,$$

where we are allowed to choose any number we wish for c_3. For example, if we choose $c_3 = 1$, then $c_1 = -2$ and $c_2 = -3$. Readers should check that these numbers satisfy equation (11.26). Consequently, $-2\mathbf{v}_1 - 3\mathbf{v}_2 + 1\mathbf{v}_3 = \mathbf{0}$ and the vectors are dependent.

The Nullspace and Dependence

A quick glance at equation (11.27) reveals that the dependence of the vectors

$$\mathbf{v}_1 = \begin{bmatrix} 1 \\ 1 \\ 1 \end{bmatrix}, \quad \mathbf{v}_2 = \begin{bmatrix} 0 \\ -1 \\ 2 \end{bmatrix}, \quad \text{and} \quad \mathbf{v}_3 = \begin{bmatrix} 2 \\ -1 \\ 8 \end{bmatrix}$$

is related to the nullspace of the matrix

$$V = \begin{bmatrix} 1 & 0 & 2 \\ 1 & -1 & -1 \\ 1 & 2 & 8 \end{bmatrix},$$

which has the vectors $\mathbf{v}_1$, $\mathbf{v}_2$, and $\mathbf{v}_3$ as its column vectors. If the nullspace of V contains a nonzero vector (as in Example 14), then the columns of V are linearly dependent, and the elements of the nonzero vector form the coefficients of a non-trivial linear combination of the columns of V which is equal to the zero vector. On the other hand, if the nullspace of V contains only the zero vector, then the columns of V are linearly independent.

What is needed is an easier method for computing a basis for the nullspace of a matrix. MATLAB's null command provides the answer.

```
>> v1 = [1;1;1]; v2 = [0;-1;2]; v3 = [2;-1;8];
>> V = [v1,v2,v3]
V =
        1              0              2
        1             -1             -1
        1              2              8
>> null(V,'r')
ans =
           -2
           -3
            1
```

MATLAB's null command, when used with the 'r' switch, computes a basis for the nullspace in a manner similar to the technique used in Example 13. You can then find elements of the nullspace by taking all possible linear combinations of the basis vectors provided by MATLAB's null command. In this case, elements of the nullspace of V are given by taking multiples of the single basis vector $[-2, -3, 1]^T$. For example, $[-4, -6, 2]$, $[-10, -15, 5]$, and $[1, 3/2, -1/2]$ are also in the nullspace[3] of V.

Let's revisit the matrix

$$A = \begin{bmatrix} 12 & -8 & -4 & 30 \\ 3 & -2 & -1 & 6 \\ 18 & -12 & -6 & 42 \\ 6 & -4 & -2 & 15 \end{bmatrix}$$

of Example 13.

```
>> A = [12 -8 -4 30;3 -2 -1 6;18 -12 -6 42;6 -4 -2 15];
>> null(A,'r')
ans =
        2/3            1/3
         1              0
         0              1
         0              0
```

Look familiar? The vectors $\mathbf{v}_1 = [2/3, 1, 0, 0]$ and $\mathbf{v}_2 = [1/3, 0, 1, 0]$ form a basis for the nullspace of matrix A. Consequently, elements of the nullspace are formed by taking linear combinations of these vectors, much as we saw in Example 13 when we wrote

$$\mathbf{x} = s \begin{bmatrix} 2/3 \\ 1 \\ 0 \\ 0 \end{bmatrix} + t \begin{bmatrix} 1/3 \\ 0 \\ 1 \\ 0 \end{bmatrix}.$$

Bases are not unique. As long as you have two independent vectors that span the nullspace of matrix A, you've got a basis. For example, it is not difficult to show that all elements of the nullspace of

---

[3] And $-4\mathbf{v}_1 - 6\mathbf{v}_2 + 2\mathbf{v}_2 = \mathbf{0}$, etc.

matrix A can also be written as linear combinations of the independent vectors $\mathbf{w}_1 = [2, 3, 0, 0]^T$ and $\mathbf{w}_2 = [1, 0, 3, 0]^T$. Also, consider the output from MATLAB's null command, this time without the 'r' switch.

```
>> format, null(A)
ans =
    -0.5976      0.0033
    -0.7196     -0.4432
    -0.3536      0.8964
    -0.0000     -0.0000
```

The command format returns the command window to the standard output format. The output of this null command is normalized so that the sum of the squares of the entries of each column vector equals one, and the dot product of different column vectors equals zero. This basis is called an *orthonormal* basis and is very useful in certain situations.

Although bases are not unique, it is true that every basis for the nullspace must contain the same number of vectors. This number is called the *dimension* of the nullspace. Thus, the nullspace of matrix A has dimension equal to 2; and, should you find another basis for the nullspace of the matrix A, it will also contain two vectors.

Example 15. *Discuss the dependence of the vectors*

$$\mathbf{v}_1 = \begin{bmatrix} 1 \\ -1 \\ 4 \end{bmatrix}, \quad \mathbf{v}_2 = \begin{bmatrix} -2 \\ 4 \\ 5 \end{bmatrix}, \quad and \quad \mathbf{v}_1 = \begin{bmatrix} -2 \\ 10 \\ 43 \end{bmatrix}.$$

```
>> v1 = [1;-1;4]; v2 = [-2;4;5]; v3 = [-2;10;43];
>> V = [v1,v2,v3]
V =
        1           -2          -2
       -1            4          10
        4            5          43
>> null(V,'r')
ans =
    Empty matrix: 3-by-0
```

This response means that there are no nonzero vectors in the nullspace, so we conclude that the vectors are linearly independent.

Exercises

In Exercises 1 – 4, find the general solution to each of the indicated system of linear equations using the method of row operations. You may use MATLAB to perform the operations, but in your submission show all of the operations that you perform. In other words, do not use rref. Use the diary command (as explained in Chapter 1) to record your work.

1. $\begin{aligned} -5x + 14y &= -47 \\ -7x + 16y &= -55 \end{aligned}$

2. $\begin{aligned} 2x_1 - 5x_2 + 3x_3 &= 8 \\ 4x_1 + 3x_2 - 7x_3 &= -3 \end{aligned}$

3. $\begin{aligned} -6x - 8y + 8z &= -30 \\ 9x + 11y - 8z &= 33 \\ 9x + 9y - 6z &= 27 \end{aligned}$

4. $\begin{aligned} -19x_1 - 128x_2 + 81x_3 + 38x_4 &= 0 \\ 8x_1 + 61x_2 - 27x_3 - 16x_4 &= 0 \\ 2x_1 + 4x_2 + 12x_3 - 4x_4 &= 0 \\ -8x_1 - 16x_2 + 12x_3 + 16x_4 &= 0 \end{aligned}$

In Exercises 5 – 10, find the general solution to the indicated system of linear equations. You may use `rref`, and the backslash operation (`A\b`) if it applies.

5. $\begin{aligned} -x - 9y + 6z &= 15 \\ 2x + 9y + z &= -16 \end{aligned}$

6. $\begin{aligned} 8x - 10y - 228z &= -112 \\ 2x - y - 4z &= -16 \\ 4x - 5y - 14z &= -56 \end{aligned}$

7. $\begin{aligned} -19x_1 - 128x_2 + 81x_3 + 38x_4 &= 3 \\ 8x_1 + 61x_2 - 27x_3 - 16x_4 &= 5 \\ 2x_1 + 4x_2 + 12x_3 - 4x_4 &= -21 \\ -8x_1 - 16x_2 + 12x_3 + 16x_4 &= -4 \end{aligned}$

8. $\begin{aligned} 2x + 2z &= 6 \\ x + z &= 3 \\ -7x + 12y + 5z &= 27 \end{aligned}$

9. $\begin{aligned} \tfrac{2}{3}x_1 - \tfrac{2}{3}x_3 + \tfrac{1}{3}x_4 &= \tfrac{1}{3} \\ \tfrac{11}{2}x_1 - 3x_2 - \tfrac{20}{3}x_3 + \tfrac{1}{3}x_4 &= -\tfrac{80}{3} \\ -3x_1 + 3x_2 + 6x_3 &= 27 \\ -\tfrac{16}{3}x_1 + 6x_2 + \tfrac{34}{3}x_3 + \tfrac{1}{3}x_4 &= \tfrac{163}{3} \end{aligned}$

10. $\begin{aligned} -23x_1 + 26x_2 - 42x_3 - 32x_4 - 90x_5 &= -6 \\ -2x_1 + x_2 - 3x_4 - 4x_5 &= -2 \\ -17x_1 + 19x_2 - 28x_3 - 22x_4 - 63x_5 &= -3 \\ -14x_1 + 14x_2 - 24x_3 - 16x_4 - 52x_5 &= -2 \\ 18x_1 - 20x_2 + 32x_3 + 23x_4 + 69x_5 &= 3 \end{aligned}$

In Exercises 11 – 16, find a basis for the nullspace of the indicated matrix. What is the dimension of the nullspace?

11. $\begin{bmatrix} 2 & 2 \\ -1 & -1 \end{bmatrix}$

12. $\begin{bmatrix} 0 & 4 & -2 \\ 1 & -2 & 1 \\ 3 & -10 & 5 \end{bmatrix}$

13. $\begin{bmatrix} 2 & -4 & -9 \\ 0 & 0 & -2 \\ 0 & 0 & 1 \end{bmatrix}$

14. $\begin{bmatrix} 12 & -5 & -14 & -9 \\ 18 & -7 & -22 & -12 \\ 12 & -6 & -12 & -9 \\ -16 & 8 & 16 & 11 \end{bmatrix}$

15. $\begin{bmatrix} -3 & 2 & 5 & 2 \\ 6 & -2 & -8 & -2 \\ -4 & 2 & 6 & 2 \\ -4 & 2 & 6 & 2 \end{bmatrix}$

16. $\begin{bmatrix} 6 & -4 & 2 & -12 & -8 \\ 6 & -4 & 2 & -12 & -8 \\ 29 & -14 & 11 & -54 & -36 \\ 13 & -6 & 5 & -24 & -16 \\ -10 & 4 & -4 & 18 & 12 \end{bmatrix}$

In Exercises 17 – 23, determine if the set of vectors is linearly dependent or independent. If they are dependent, find a nonzero linear combination which is equal to the zero vector.

17. $\begin{bmatrix} 1 \\ 2 \end{bmatrix}$ and $\begin{bmatrix} -3 \\ -6 \end{bmatrix}$

18. $\begin{bmatrix} 1 \\ 2 \end{bmatrix}$ and $\begin{bmatrix} -3 \\ -5 \end{bmatrix}$

19. $\begin{bmatrix} 1 \\ 2 \\ 0 \end{bmatrix}$ and $\begin{bmatrix} -3 \\ -6 \\ 1 \end{bmatrix}$

20. $\begin{bmatrix} 1 \\ 1 \\ 1 \end{bmatrix}$, $\begin{bmatrix} 1 \\ -1 \\ 1 \end{bmatrix}$, and $\begin{bmatrix} 5 \\ 0 \\ 5 \end{bmatrix}$

21. $\begin{bmatrix} 1 \\ 1 \\ 1 \end{bmatrix}$, $\begin{bmatrix} 1 \\ -1 \\ 1 \end{bmatrix}$, and $\begin{bmatrix} 5 \\ 1 \\ 5 \end{bmatrix}$

22. $\begin{bmatrix} 1 \\ 0 \\ 1 \\ 0 \end{bmatrix}$, $\begin{bmatrix} 0 \\ 1 \\ 1 \\ 1 \end{bmatrix}$, and $\begin{bmatrix} 5 \\ -6 \\ -1 \\ -6 \end{bmatrix}$

23. $\begin{bmatrix} 1 \\ 0 \\ 1 \\ 0 \end{bmatrix}$, $\begin{bmatrix} 0 \\ 1 \\ 1 \\ 1 \end{bmatrix}$, and $\begin{bmatrix} 5 \\ -6 \\ 0 \\ -6 \end{bmatrix}$

12.   Homogeneous Linear Systems of ODEs

A system of first order differential equations is *linear*, *homogeneous*, and has *constant coefficients* if it has the specific form

$$
\begin{aligned}
x_1{}' &= a_{11}x_1 + a_{12}x_2 + \cdots + a_{1n}x_n, \\
x_2{}' &= a_{21}x_1 + a_{22}x_2 + \cdots + a_{2n}x_n, \\
&\;\;\vdots \\
x_n{}' &= a_{n1}x_1 + a_{n2}x_2 + \cdots + a_{nn}x_n,
\end{aligned}
\tag{12.1}
$$

where each a_{ij} is a constant. System (12.1) can be written in matrix form

$$
\mathbf{x}' =
\begin{bmatrix} x_1 \\ x_2 \\ \vdots \\ x_n \end{bmatrix}'
=
\begin{bmatrix}
a_{11} & a_{12} & \cdots & a_{1n} \\
a_{21} & a_{22} & \cdots & a_{2n} \\
\vdots & \vdots & \ddots & \vdots \\
a_{n1} & a_{n2} & \cdots & a_{nn}
\end{bmatrix}
\begin{bmatrix} x_1 \\ x_2 \\ \vdots \\ x_n \end{bmatrix}
= A\mathbf{x}.
\tag{12.2}
$$

The equation $\mathbf{x}' = A\mathbf{x}$ in (12.2) is strikingly similar to the simple first order differential equation

$$
x' = \lambda x.
\tag{12.3}
$$

It is easy to check that

$$
x = Ce^{\lambda t}
$$

is a solution to equation (12.3), where C is an arbitrary constant. Consequently, it seems natural that we should first try to find a solution of equation $\mathbf{x}' = A\mathbf{x}$ having the form

$$
\mathbf{x} = e^{\lambda t}\mathbf{v},
\tag{12.4}
$$

where $\mathbf{v}$ is a column n-vector of constants. The constant function $\mathbf{x} = \mathbf{0}$ is a solution of equation (12.2), but not a very interesting one. Let's agree that we want to find non-zero solutions of equation (12.2). That is, we wish to find a solution of the form $\mathbf{x} = e^{\lambda t}\mathbf{v}$, where $\mathbf{v}$ is a non-zero vector of constants (some of the entries in $\mathbf{v}$ may be zero, but not all). If we substitute equation (12.4) into the left- and right-hand sides of the equation $\mathbf{x}' = A\mathbf{x}$, we get

$$
\mathbf{x}' = (e^{\lambda t}\mathbf{v})' = \lambda e^{\lambda t}\mathbf{v} \quad \text{and}
$$
$$
A\mathbf{x} = A\left(e^{\lambda t}\mathbf{v}\right) = e^{\lambda t}A\mathbf{v}.
$$

For these to be equal we must have $\lambda e^{\lambda t}\mathbf{v} = e^{\lambda t}A\mathbf{v}$. If we divide both sides of this equation by $e^{\lambda t}$, which is non-zero, we arrive at

$$
A\mathbf{v} = \lambda\mathbf{v}.
\tag{12.5}
$$

The problem of finding a solution of equation (12.2) is now reduced to finding a number λ and a non-zero vector of constants $\mathbf{v}$ that satisfy the equation $A\mathbf{v} = \lambda\mathbf{v}$.

Eigenvalues Using MATLAB

Solutions of the equation $A\mathbf{v} = \lambda\mathbf{v}$ have far reaching application in many areas of mathematics and science. In particular they will enable us to find the solutions of equation (12.2). Let's make a formal definition.

Definition. *Let A be an $n \times n$ matrix. A number λ is said to be an eigenvalue for A if there is a non-zero vector $\mathbf{v}$ such that $A\mathbf{v} = \lambda\mathbf{v}$. If λ is an eigenvalue for A, then any vector $\mathbf{v}$ which satisfies $A\mathbf{v} = \lambda\mathbf{v}$ is called an eigenvector for A associated to the eigenvalue λ.*

Let I be the $n \times n$ identity matrix, which has ones along the main diagonal and zeros elsewhere. Then $I\mathbf{v} = \mathbf{v}$ for every column n-vector $\mathbf{v}$. This allows us to manipulate the equation $A\mathbf{v} = \lambda\mathbf{v}$ as follows:

$$\begin{aligned} \mathbf{0} &= A\mathbf{v} - \lambda\mathbf{v}, \\ &= A\mathbf{v} - \lambda I\mathbf{v}, \\ &= (A - \lambda I)\mathbf{v}. \end{aligned} \qquad (12.6)$$

We now apply Theorem 2 of Chapter 11. If $\det(A - \lambda I) \neq 0$, then equation (12.6) has the *unique* solution $\mathbf{v} = \mathbf{0}$. However, we are looking for *non-zero* solutions of equation (12.6), so we conclude that λ is an eigenvalue of A if and only if $\det(A - \lambda I) = 0$.

Definition. *Let A be an $n \times n$ matrix. The function*

$$p(\lambda) = (-1)^n \det(A - \lambda I) = \det(\lambda I - A)$$

is called the **characteristic polynomial** *of the matrix A. The zeros of the characteristic polynomial are the eigenvalues of the matrix A.*

The factor of $(-1)^n$ is chosen to make the leading term in the characteristic polynomial equal to λ^n instead of $(-1)^n \lambda^n$.[1]

Example 1. *Find the eigenvalues of the matrix*

$$A = \begin{bmatrix} -3 & 1 & -3 \\ -8 & 3 & -6 \\ 2 & -1 & 2 \end{bmatrix}.$$

---

[1] This definition may not agree with what you find in your textbook. Some authors choose to define the characteristic polynomial as $\det(A - \lambda I)$, which differs by the factor of $(-1)^n$ from our definition. Some authors of more than one book even use different definitions in different books. In fact the two definitions differ only by a factor of ± 1. Since we are only interested in the zeros of the characteristic polynomial, this makes little difference. We are choosing the defintion that agrees with the usage in MATLAB.

We start by calculating the characteristic polynomial. Notice that $n = 3$ in this case, so

$$p(\lambda) = (-1)^3 \det(A - \lambda I),$$

$$= -\det\left(\begin{bmatrix} -3 & 1 & -3 \\ -8 & 3 & -6 \\ 2 & -1 & 2 \end{bmatrix} - \lambda \begin{bmatrix} 1 & 0 & 0 \\ 0 & 1 & 0 \\ 0 & 0 & 1 \end{bmatrix}\right),$$

$$= -\det\begin{bmatrix} -3-\lambda & 1 & -3 \\ -8 & 3-\lambda & -6 \\ 2 & -1 & 2-\lambda \end{bmatrix},$$

$$= \lambda^3 - 2\lambda^2 - \lambda + 2.$$

Of course, this last step requires that we are pretty good at calculating the determinant of a rather complicated matrix. MATLAB can help you with the computation.

```
>> A = [-3 1 -3;-8 3 -6;2 -1 2];
>> p = poly(A)
p =
      1.0000   -2.0000   -1.0000    2.0000
```

Note that the vector p = [1 -2 -1 2] contains the coefficients of the characteristic polynomial $p(\lambda) = \lambda^3 - 2\lambda^2 - \lambda + 2$. To find the eigenvalues we must find the zeros of the characteristic polynomial. If you are good at factoring you will discover that

$$p(\lambda) = \lambda^3 - 2\lambda^2 - \lambda + 2 = (\lambda - 1)(\lambda + 1)(\lambda - 2).$$

Thus, the eigenvalues of A are $\lambda = 1, -1$, and 2. MATLAB can help you find the roots. If the vector p contains the coefficients of the characteristic polynomial, then we can use roots(p) to compute the zeros of p.

```
>> roots(p)
ans =
    -1.0000
     2.0000
     1.0000
```

Symbolic Toolbox Users. If you change your matrix into a symbolic object, Maple can be called upon to perform the evaluations. For example,

```
>> ps = poly(sym(A))
ps =
x^3-2*x^2-x+2
```

computes the characteristic polynomial as a mathematical expression. You can then factor the polynomial with the command

```
>> factor(ps)
ans =
(x-1)*(x-2)*(x+1)
```

It is clear that the zeros are 1, 2, and -1, but you can also use the Symbolic Toolbox's `solve` command.

```
>> solve(ps)
ans =
   [ -1]
   [  1]
   [  2]
```

Eigenvectors Using MATLAB

Having found the eigenvalues, we must now find the associated eigenvectors. Let's look again at equation (12.6), which we now write as

$$(A - \lambda I)\mathbf{v} = \mathbf{0}. \tag{12.7}$$

This is the equation which must be solved by an eigenvalue λ and an associated eigenvector $\mathbf{v}$. This means that the eigenvector $\mathbf{v}$ is an element of the nullspace of the matrix $A - \lambda I$.

Example 2. *Find the eigenvectors of the matrix*

$$A = \begin{bmatrix} -3 & 1 & -3 \\ -8 & 3 & -6 \\ 2 & -1 & 2 \end{bmatrix}.$$

The eigenvalues of the matrix A were found in Example 1 to be $\lambda = 1$, -1, and 2. Select one of the eigenvalues, say $\lambda_1 = 1$. The eigenvectors associated with $\lambda_1 = 1$ are the vectors in the nullspace of $A - \lambda_1 I = A - 1I$. These can be found using the methods of Chapter 11. We could use row operations, but we shall use the command `null`. (Notice that the 3×3 identity matrix I is generated with the MATLAB command `eye(3)`.)

```
>> v1 = null(A - 1*eye(3),'r')
v1 =
      -1
      -1
       1
```

The vectors in the nullspace of $A - 1I$ are the multiples of $\mathbf{v}_1$, and they are all eigenvectors. Any nonzero multiple of $\mathbf{v}_1$ will serve our purpose, so let's choose $\mathbf{v}_1$ itself.

In a similar manner, the MATLAB commands

```
>> v2 = null(A - (-1)*eye(3),'r')
v2 =
   0.5000
   1.0000
        0
```

184

```
>> v3 = null(A - 2*eye(3),'r')
v3 =
    -1
    -2
     1
```

provide eigenvectors $\mathbf{v}_2 = [1/2, 1, 0]^T$ and $\mathbf{v}_3 = [-1, -2, 1]^T$ associated with the eigenvalues $\lambda_2 = -1$ and $\lambda_3 = 2$, respectively. Any nonzero multiples of these vectors would also be suitable[2]. For example, we might want our vectors to contain only integer entries, in which case we could choose $\mathbf{v}_2 = [1, 2, 0]^T$ for the eigenvalue $\lambda_2 = -1$.

Symbolic Toolbox Users. The `null` command works equally well with symbolic objects, but the `'r'` switch is not needed.

```
>> v2 = null(sym(A) - (-1)*eye(3))
v2 =
[ 1]
[ 2]
[ 0]
```

Finally, you can easily check your solution, symbolically (with the Symbolic Toolbox) or numerically (with MATLAB). For example, use the Symbolic Toolbox to show that $A\mathbf{v}_2 = -1\mathbf{v}_2$.

```
>> A*v2, -1*v2
ans =
[ -1]
[ -2]
[  0]
ans =
[ -1]
[ -2]
[  0]
```

Since $A\mathbf{v}_2 = -1\mathbf{v}_2$, $\mathbf{v}_2$ is an eigenvector with associated eigenvalue $\lambda_2 = -1$.

MATLAB's eig Command

Finding eigenvalues and eigenvectors in the manner described requires three steps. First, find the characteristic polynomial; second, find the eigenvalues, which are the roots of the characteristic polynomial; and third, for each eigenvalue λ, find the associated eigenvectors by finding the nullspace of $A - \lambda I$. This three step process can be replaced with a one step process using MATLAB's `eig` command.

Example 3. *Use MATLAB's eig command to find the eigenvalues and eigenvectors of*

$$A = \begin{bmatrix} -3 & 1 & -3 \\ -8 & 3 & -6 \\ 2 & -1 & 2 \end{bmatrix}.$$

---

[2] If $\mathbf{v}$ is an eigenvector of matrix A associated with the eigenvalue λ, then $A(c\mathbf{v}) = c(A\mathbf{v}) = c(\lambda\mathbf{v}) = \lambda(c\mathbf{v})$, making $c\mathbf{v}$ an eigenvector for any constant number c.

The command `eig(A)` will display the eigenvalues of A.

```
>> A = [-3 1 -3;-8 3 -6;2 -1 2];
>> eig(A)
ans =
   -1.0000
    2.0000
    1.0000
```

However, the command `[V,E] = eig(A)` will output two matrices. E will be a diagonal matrix with the eigenvalues along the diagonal, and V will have the associated eigenvectors as its column vectors.

```
>> [V,E] = eig(A)
V =
    0.4472   -0.4082   -0.5774
    0.8944   -0.8165   -0.5774
   -0.0000    0.4082    0.5774
E =
   -1.0000        0        0
        0   2.0000        0
        0        0   1.0000
```

In this representation the eigenvector associated to an eigenvalue is the column vector directly above the eigenvalue. For example, the eigenvector associated with the eigenvalue 2 is the second column of V.

We have made the point several times that eigenvectors are not unique. Indeed the eigenvectors found using `eig` are not the same as those we found in Example 2. However, the eigenvectors found by the `eig` command must be scalar multiples of those found in Example 2. The command

```
>> V(:,2)/V(3,2)
ans =
   -1.0000
   -2.0000
    1.0000
```

shows that the second column in V is a multiple of the eigenvector v_3 we found in Example 2. We will leave it to you to show that the same is true for the other cases.

The eigenvectors that are produced by `eig` are normalized so that the sum of the squares of the elements of each eigenvector is equal to 1. This means that the eigenvectors have geometric length 1. For example,

```
>> norm(V(:,2))
ans =
    1
```

Symbolic Toolbox Users. MATLAB's eig command will access Maple if the matrix A is a symbolic object.

```
>> [V,E] = eig(sym(A))
V =
[  1, -1, -1]
[  2, -2, -1]
[  0,  1,  1]
E =
[ -1,  0,  0]
[  0,  2,  0]
[  0,  0,  1]
```

The symbolic version of eig uses methods similar to those we used in Chapter 11, so it is not surprising that the answers are similar to those found in Example 2.

Tying It All Together—Solving Systems of ODEs

We are now ready to solve homogeneous, linear systems with constant coefficients.

Example 4. *Solve the the initial value problem*

$$x_1' = 8x_1 - 5x_2 + 10x_3,$$
$$x_2' = 2x_1 + x_2 + 2x_3, \qquad (12.8)$$
$$x_3' = -4x_1 + 4x_2 - 6x_3,$$

where $x_1(0) = 2$, $x_2(0) = 2$, and $x_3(0) = -3$.

We start by writing the system as the matrix equation

$$\mathbf{x}' = \begin{bmatrix} x_1 \\ x_2 \\ x_3 \end{bmatrix}' = \begin{bmatrix} 8 & -5 & 10 \\ 2 & 1 & 2 \\ -4 & 4 & -6 \end{bmatrix} \begin{bmatrix} x_1 \\ x_2 \\ x_3 \end{bmatrix} = A\mathbf{x}.$$

Next we execute

```
>> A = [8,-5,10;2,1,2;-4,4,-6];
>> eig(A)
```

to enter the matrix A and compute that its eigenvalues are $\lambda_1 = -2$, $\lambda_2 = 3$, and $\lambda_3 = 2$. Next we execute >> format rat to usecompute

```
>> v1 = null(A - (-2)*eye(3),'r')
>> v2 = null(A - (3)*eye(3),'r')
>> v3 = null(A - (2)*eye(3),'r')
```

and find that the eigenvectors are

$$\mathbf{v}_1 = \begin{bmatrix} -1 \\ 0 \\ 1 \end{bmatrix}, \quad \mathbf{v}_2 = \begin{bmatrix} 1 \\ 1 \\ 0 \end{bmatrix}, \quad \text{and} \quad \mathbf{v}_3 = \begin{bmatrix} 0 \\ 2 \\ 1 \end{bmatrix}.$$

The theory developed at the beginning of the chapter states that each eigenvalue-eigenvector pair $(\lambda, \mathbf{v})$ produces the solution $\mathbf{x} = e^{\lambda t}\mathbf{v}$ of $\mathbf{x}' = A\mathbf{x}$. In our case we get the three solutions

$$\mathbf{x}_1 = e^{-2t}\begin{bmatrix} -1 \\ 0 \\ 1 \end{bmatrix}, \quad \mathbf{x}_2 = e^{3t}\begin{bmatrix} 1 \\ 1 \\ 0 \end{bmatrix}, \quad \text{and} \quad \mathbf{x}_3 = e^{2t}\begin{bmatrix} 0 \\ 2 \\ 1 \end{bmatrix}.$$

According to the general theory of linear systems we are through. Linear combinations of solutions to a linear system are also solutions. The general solution of an $n \times n$ system is given by the general linear combination of n linearly independent solutions, called a *fundamental system of solutions*. Furthermore, solutions corresponding to distinct eigenvalues are automatically linearly independent. Thus, for our 3×3 system the solutions $\mathbf{x}_1$, $\mathbf{x}_2$, and $\mathbf{x}_3$ are linearly independent and form a fundamental system of solutions. The general solution of the system is the general linear combination of $\mathbf{x}_1$, $\mathbf{x}_2$, and $\mathbf{x}_3$,

$$\mathbf{x}(t) = c_1\mathbf{x}_1(t) + c_2\mathbf{x}_2(t) + c_3\mathbf{x}_3(t) = c_1 e^{-2t}\begin{bmatrix} -1 \\ 0 \\ 1 \end{bmatrix} + c_2 e^{3t}\begin{bmatrix} 1 \\ 1 \\ 0 \end{bmatrix} + c_3 e^{2t}\begin{bmatrix} 0 \\ 2 \\ 1 \end{bmatrix}. \tag{12.9}$$

We want the particular solution that satisfies the initial condition

$$\mathbf{x}(0) = \begin{bmatrix} x_1(0) \\ x_2(0) \\ x_3(0) \end{bmatrix} = \begin{bmatrix} 2 \\ 2 \\ -3 \end{bmatrix}.$$

Substituting $t = 0$ in equation (12.9) yields

$$\mathbf{x}(0) = c_1 e^{-2(0)}\begin{bmatrix} -1 \\ 0 \\ 1 \end{bmatrix} + c_2 e^{3(0)}\begin{bmatrix} 1 \\ 1 \\ 0 \end{bmatrix} + c_3 e^{2(0)}\begin{bmatrix} 0 \\ 2 \\ 1 \end{bmatrix}, \quad \text{or}$$

$$\begin{bmatrix} 2 \\ 2 \\ -3 \end{bmatrix} = c_1\begin{bmatrix} -1 \\ 0 \\ 1 \end{bmatrix} + c_2\begin{bmatrix} 1 \\ 1 \\ 0 \end{bmatrix} + c_3\begin{bmatrix} 0 \\ 2 \\ 1 \end{bmatrix}. \tag{12.10}$$

Equation (12.10) is equivalent to the matrix equation

$$\begin{bmatrix} -1 & 1 & 0 \\ 0 & 1 & 2 \\ 1 & 0 & 1 \end{bmatrix}\begin{bmatrix} c_1 \\ c_2 \\ c_3 \end{bmatrix} = \begin{bmatrix} 2 \\ 2 \\ -3 \end{bmatrix},$$

an equation in the form $V\mathbf{c} = \mathbf{b}$, where the column vectors of the matrix V are the eigenvectors $\mathbf{v}_1$, $\mathbf{v}_2$, and $\mathbf{v}_3$. The solution is easily found with the following MATLAB commands.

```
>> V = [v1,v2,v3], b = [2;2;-3]
V =
        -1              1              0
         0              1              2
         1              0              1
b =
         2
         2
        -3
>> c = V\b
c =
        -6
        -4
         3
```

188

Thus, $c_1 = -6$, $c_2 = -4$, $c_3 = 3$, and the solution of the initial value problem (12.8) is

$$\mathbf{x}(t) = -6e^{-2t} \begin{bmatrix} -1 \\ 0 \\ 1 \end{bmatrix} - 4e^{3t} \begin{bmatrix} 1 \\ 1 \\ 0 \end{bmatrix} + 3e^{2t} \begin{bmatrix} 0 \\ 2 \\ 1 \end{bmatrix}.$$

Finally, since $\mathbf{x}(t) = [x_1(t), x_2(t), x_3(t)]^T$, we can write the solution of system (12.8) as

$$\mathbf{x}(t) = \begin{bmatrix} x_1(t) \\ x_2(t) \\ x_3(t) \end{bmatrix} = \begin{bmatrix} 6e^{-2t} - 4e^{3t} \\ -4e^{3t} + 6e^{2t} \\ -6e^{-2t} + 3e^{2t} \end{bmatrix}.$$

Example 5. *Solve the initial value problem*

$$\begin{aligned} x_1' &= -9x_1 + 4x_2, \\ x_2' &= -2x_1 + 2x_2, \end{aligned} \tag{12.11}$$

with initial values $x_1(0) = 1$ and $x_2(0) = -1$.

Set up the system as the matrix equation.

$$\mathbf{x}' = \begin{bmatrix} x_1 \\ x_2 \end{bmatrix}' = \begin{bmatrix} -9 & 4 \\ -2 & 2 \end{bmatrix} \begin{bmatrix} x_1 \\ x_2 \end{bmatrix} = A\mathbf{x}.$$

The characteristic polynomial of matrix A is $p(\lambda) = \lambda^2 + 7\lambda - 10$, whose zeros are $-7/2 \pm (1/2)\sqrt{89}$. We can give exact values of the eigenvalues and eigenvectors if we are willing to use expressions involving square roots. Instead, let's use the decimal approximations given by MATLAB's `eig` command.

```
>> A = [-9 4;-2 2];
>> [V,E] = eig(A)
V =
    -0.9814    -0.3646
    -0.1921    -0.9312
E =
    -8.2170         0
         0    1.2170
```

Since scalar multiples of eigenvectors remain eigenvectors, we will still have eigenvectors if we replace V with −V. (Note: we could easily have kept the original eigenvectors. There just were too many negative signs for our taste.)

```
>> V = -V
V =
     0.9814     0.3646
     0.1921     0.9312
```

These results easily lead to the general solution.

$$\mathbf{x}(t) = c_1 e^{-8.2170t} \begin{bmatrix} 0.9814 \\ 0.1921 \end{bmatrix} + c_2 e^{1.2170t} \begin{bmatrix} 0.3646 \\ 0.9312 \end{bmatrix}. \tag{12.12}$$

We need to determine the solution that satisfies the initial condition

$$\mathbf{x}(0) = \begin{bmatrix} x_1(0) \\ x_2(0) \end{bmatrix} = \begin{bmatrix} 1 \\ -1 \end{bmatrix}.$$

Substituting $t = 0$ in equation (12.12) yields

$$\begin{bmatrix} 1 \\ -1 \end{bmatrix} = c_1 \begin{bmatrix} 0.9814 \\ 0.1921 \end{bmatrix} + c_2 \begin{bmatrix} 0.3646 \\ 0.9312 \end{bmatrix},$$

which is equivalent to the matrix equation

$$\begin{bmatrix} 0.9814 & 0.3646 \\ 0.1921 & 0.9312 \end{bmatrix} \begin{bmatrix} c_1 \\ c_2 \end{bmatrix} = \begin{bmatrix} 1 \\ -1 \end{bmatrix}.$$

This is an equation in the form $V\mathbf{c} = \mathbf{b}$, and is easily solved in MATLAB.

```
>> b = [1;-1];
>> c = V\b
c =
    1.5356
   -1.3907
```

Therefore, the solution of the initial value problem for system (12.11) is

$$\mathbf{x}(t) = 1.5356e^{-8.2170t} \begin{bmatrix} 0.9814 \\ 0.1921 \end{bmatrix} - 1.3907e^{1.2170t} \begin{bmatrix} 0.3646 \\ 0.9312 \end{bmatrix}.$$

Example 6. *Find the general solution of the system*

$$x_1' = 15x_1 - 6x_2 - 18x_3 - 6x_4,$$
$$x_2' = -4x_1 + 5x_2 + 8x_3 + 4x_4,$$
$$x_3' = 12x_1 - 6x_2 - 15x_3 - 6x_4,$$
$$x_4' = 4x_1 - 2x_2 - 8x_3 - x_4.$$

Set up the system as the matrix equation,

$$\mathbf{x}' = \begin{bmatrix} x_1 \\ x_2 \\ x_3 \\ x_4 \end{bmatrix}' = \begin{bmatrix} 15 & -6 & -18 & -6 \\ -4 & 5 & 8 & 4 \\ 12 & -6 & -15 & -6 \\ 4 & -2 & -8 & -1 \end{bmatrix} \begin{bmatrix} x_1 \\ x_2 \\ x_3 \\ x_4 \end{bmatrix} = A\mathbf{x}.$$

Enter the matrix A into MATLAB's workspace and compute the eigenvalues,

```
>> A = [15 -6 -18 -6;-4 5 8 4;12 -6 -15 -6;4 -2 -8 -1];
>> E = eig(A)
E =
   -3.0000
    3.0000
    1.0000
    3.0000
```

190

What sets this matrix apart from previous matrices is the presence of a *repeated* eigenvalue. The eigenvalue 3 is an eigenvalue of multiplicity two[3].

The eigenvectors associated with $\lambda_1 = -3$ and $\lambda_2 = 1$ are easily captured with MATLAB's null command.

```
>> v1 = null(A - (-3)*eye(4),'r')
v1 =
        1
       -1
        1
        1


>> v2 = null(A - 1*eye(4),'r')
v2 =
        0
       -1
        0
        1
```

The three equations in Example 4 required three independent eigenvectors to form the general solution. The two equations in Example 5 required two independent eigenvectors to form the general solution. In this example we have four equations, so four independent eigenvectors are required. We have two independent eigenvectors, but only one eigenvalue remains. Let's cross our fingers for luck and use MATLAB's null command once more.

```
>> null(A - 3*eye(4),'r')
ans =
    0.5000   -1.0000
    1.0000        0
         0   -1.0000
         0    1.0000
```

Fortunately for us[4], the eigenspace associated with the eigenvalue $\lambda_3 = 3$ has two basis vectors, exactly the number needed to produce the general solution

$$\mathbf{x}(t) = c_1 e^{-3t} \begin{bmatrix} 1 \\ -1 \\ 1 \\ 1 \end{bmatrix} + c_2 e^{1t} \begin{bmatrix} 0 \\ -1 \\ 0 \\ 1 \end{bmatrix} + c_3 e^{3t} \begin{bmatrix} 1/2 \\ 1 \\ 0 \\ 0 \end{bmatrix} + c_4 e^{3t} \begin{bmatrix} -1 \\ 0 \\ -1 \\ 1 \end{bmatrix}.$$

---

[3] Symbolic Toolbox users might find the sequence of commands p = poly(sym(A)); factor(p) illuminating. Does this demonstrate why 3 is an eigenvalue of multiplicity two?

[4] In later sections of this chapter we will show how to form the general solution when you have fewer than the required number of eigenvectors.

Complex Eigenvalues

Two key facts aid in the solution of systems involving complex eigenvalues and eigenvectors. Foremost is Euler's famous identity,

$$e^{i\theta} = \cos\theta + i\sin\theta.$$

Secondly, if you find a complex solution of a linear, homogeneous system with constant *real* coefficients, then both the real and imaginary parts of your complex solution will also be solutions of your system.

Example 7. *Solve the initial value problem*

$$x_1' = -x_1 + 2x_3,$$
$$x_2' = 2x_1 + 3x_2 - 6x_3, \qquad\qquad (12.13)$$
$$x_3' = -2x_1 - x_3,$$

where $x_1(0) = 4$, $x_2(0) = -5$, and $x_3(0) = 9$.

Set up the system as the matrix equation

$$\mathbf{x}' = \begin{bmatrix} x_1 \\ x_2 \\ x_3 \end{bmatrix}' = \begin{bmatrix} -1 & 0 & 2 \\ 2 & 3 & -6 \\ -2 & 0 & -1 \end{bmatrix}\begin{bmatrix} x_1 \\ x_2 \\ x_3 \end{bmatrix} = A\mathbf{x}.$$

Note that the entries of the matrix A are real numbers, so the coefficients of the characteristic polynomial must be real numbers, which is easily checked in MATLAB.

```
>> A = [-1 0 2;2 3 -6;-2 0 -1];
>> p = poly(A)
p =
    1.0000   -1.0000   -1.0000   -15.0000
```

If a polynomial has real coefficients, any complex zeros must occur in complex conjugate[5] pairs.

```
>> r = roots(p)
r =
    3.0000
   -1.0000 + 2.0000i
   -1.0000 - 2.0000i
```

Note that both $-1+2i$ and $-1-2i$ are eigenvalues; the eigenvalues occur in conjugate pairs. Furthermore, the eigenvectors must also occur in complex conjugate pairs[6]. The computation of the eigenvalues of A can be confirmed by executing

```
>> eig(A)
ans =
    3.0000
   -1.0000 + 2.0000i
   -1.0000 - 2.0000i
```

---

[5] The conjugate of $a + bi$ is $a - bi$.

[6] The conjugate of a vector or matrix is found by taking the complex conjugate of each entry. For example, the conjugate of $[1/2, -1/2 + (1/2)i, (1/2)i]^T$ is $[1/2, -1/2 - (1/2)i, -(1/2)i]^T$.

We can form the general solution if we can find three linearly independent solutions. We find an eigenvector associated with the eigenvalue 3 by executing

```
>> v1 = null(A - 3*eye(3),'r')
v1 =
     0
     1
     0
```

Therefore the function

$$\mathbf{x}_1(t) = e^{3t} \begin{bmatrix} 0 \\ 1 \\ 0 \end{bmatrix}$$

is a solution of system (12.13).

When we find an eigenvector associated with the complex eigenvalue $-1 + 2i$, we get

```
>> w = null(A - (-1+2i)*eye(3),'r')
w =
        0 - 1.0000i
   1.0000 + 1.0000i
   1.0000
```

Therefore, a complex solution of system (12.13) is

$$\mathbf{z}(t) = e^{(-1+2i)t} \begin{bmatrix} -i \\ 1+i \\ 1 \end{bmatrix} = e^{-t}e^{i2t} \left(\begin{bmatrix} 0 \\ 1 \\ 1 \end{bmatrix} + i \begin{bmatrix} -1 \\ 1 \\ 0 \end{bmatrix} \right).$$

Euler's formula and a little complex arithmetic reveals that

$$\mathbf{z}(t) = e^{-t}(\cos 2t + i \sin 2t) \left(\begin{bmatrix} 0 \\ 1 \\ 1 \end{bmatrix} + i \begin{bmatrix} -1 \\ 1 \\ 0 \end{bmatrix} \right)$$

$$= \left(e^{-t} \begin{bmatrix} \sin 2t \\ \cos 2t - \sin 2t \\ \cos 2t \end{bmatrix} + ie^{-t} \begin{bmatrix} -\cos 2t \\ \cos 2t + \sin 2t \\ \sin 2t \end{bmatrix} \right).$$

Now we use the fact that the real and imaginary parts of $\mathbf{z}$ form real, independent solutions of system (12.13) to get the solutions

$$\mathbf{x}_2(t) = e^{-t} \begin{bmatrix} \sin 2t \\ \cos 2t - \sin 2t \\ \cos 2t \end{bmatrix} \quad \text{and} \quad \mathbf{x}_3(t) = e^{-t} \begin{bmatrix} -\cos 2t \\ \cos 2t + \sin 2t \\ \sin 2t \end{bmatrix}$$

We now have three independent solutions $\mathbf{x}_1(t)$, $\mathbf{x}_2(t)$, and $\mathbf{x}_3(t)$, so we can form the general solution

$$\mathbf{x}(t) = c_1\mathbf{x}_1(t) + c_2\mathbf{x}_2(t) + c_3\mathbf{x}_3(t)$$

$$= c_1e^{3t} \begin{bmatrix} 0 \\ 1 \\ 0 \end{bmatrix} + c_2e^{-t} \begin{bmatrix} \sin 2t \\ \cos 2t - \sin 2t \\ \cos 2t \end{bmatrix} + c_3e^{-t} \cdot \begin{bmatrix} -\cos 2t \\ \cos 2t + \sin 2t \\ \sin 2t \end{bmatrix} \quad (12.14)$$

We want the solution that satisfies the initial condition $\mathbf{x}(0) = [x_1(0), x_2(0), x_3(0)]^T = [4, -5, 9]^T$. Substitute $t = 0$ in equation (12.14) to get

$$\begin{bmatrix} 4 \\ -5 \\ 9 \end{bmatrix} = c_1 \begin{bmatrix} 0 \\ 1 \\ 0 \end{bmatrix} + c_2 \begin{bmatrix} 0 \\ 1 \\ 1 \end{bmatrix} + c_3 \begin{bmatrix} -1 \\ 1 \\ 0 \end{bmatrix} = \begin{bmatrix} 0 & 0 & -1 \\ 1 & 1 & 1 \\ 0 & 1 & 0 \end{bmatrix} \begin{bmatrix} c_1 \\ c_2 \\ c_3 \end{bmatrix},$$

a matrix equation in the form $W\mathbf{c} = \mathbf{b}$. We can solve using the MATLAB commands

```
>> W = [0 0 -1;1 1 1;0 1 0];
>> b = [4;-5;9];
>> c = W\b
c =
    -10
      9
     -4
```

Substitution of $c_1 = -10$, $c_2 = 9$, and $c_3 = -4$ in equation (12.14), a little arithmetic, and we arrive at the final solution of the initial value problem posed in equation (12.13).

$$\begin{bmatrix} x_1(t) \\ x_2(t) \\ x_3(t) \end{bmatrix} = \begin{bmatrix} e^{-t}(4\cos 2t + 9\sin 2t) \\ -10e^{3t} + e^{-t}(5\cos 2t - 13\sin 2t) \\ e^{-t}(9\cos 2t - 4\sin 2t) \end{bmatrix}$$

The Exponential of a Matrix

In order to solve systems which do not have a full complement of eigenvectors, we need a new idea. It is provided by the exponential of a matrix.

Definition. *If A is a $n \times n$ matrix, then the exponential of A is defined by*

$$e^A = I + A + \frac{A^2}{2!} + \frac{A^3}{3!} + \cdots + \frac{A^k}{k!} + \cdots, \tag{12.15}$$

where I is the $n \times n$ identity matrix.

It can be shown that the infinite series (12.15) converges for every $n \times n$ matrix A. In all but a few simple cases, it's quite difficult to compute the exponential of a matrix with this definition. However, diagonal matrices are always easy to work with. For example, if

$$D = \begin{bmatrix} a & 0 \\ 0 & b \end{bmatrix},$$

we can compute

$$D^2 = \begin{bmatrix} a^2 & 0 \\ 0 & b^2 \end{bmatrix}, \quad D^3 = \begin{bmatrix} a^3 & 0 \\ 0 & b^3 \end{bmatrix}, \quad \cdots \quad , D^n = \begin{bmatrix} a^n & 0 \\ 0 & b^n \end{bmatrix}.$$

Hence,

$$e^D = I + D + \frac{1}{2!}D^2 + \cdots,$$

$$= \begin{bmatrix} 1 & 0 \\ 0 & 1 \end{bmatrix} + \begin{bmatrix} a & 0 \\ 0 & b \end{bmatrix} + \frac{1}{2!}\begin{bmatrix} a^2 & 0 \\ 0 & b^2 \end{bmatrix} + \cdots$$

$$= \begin{bmatrix} 1 + a + \frac{1}{2!}a^2 + \cdots & 0 \\ 0 & 1 + b + \frac{1}{2!}b^2 + \cdots \end{bmatrix}$$

$$= \begin{bmatrix} e^a & 0 \\ 0 & e^b \end{bmatrix}$$

While this calculation was for a 2×2 diagonal matrix, a similar calculation demonstrates that the exponential of a diagonal matrix of any size is equal to the diagonal matrix found by exponentiating the diagonal entries of the original matrix.

Example 8. *If I is the $n \times n$ identity matrix and r is a scalar (number), prove that $e^{rI} = e^r I$.*

Since rI is the diagonal matrix with all diagonal entries equal to r, the result is immediate from the previous computation. For example,

$$e^{3I} = e^3 I = e^3 \begin{bmatrix} 1 & 0 \\ 0 & 1 \end{bmatrix} = \begin{bmatrix} e^3 & 0 \\ 0 & e^3 \end{bmatrix}.$$

You can easily check this result with MATLAB's expm command[7]. Note that $e^3 \approx 20.0855$.

```
>> I = eye(2);
>> E = expm(3*I)
E =
    20.0855         0
         0    20.0855
```

Properties of the exponential of a matrix. The most important property of the exponential function is the formula $e^{a+b} = e^a e^b$. We should ask if this formula remains true for the exponential of a matrix.

Proposition 1. *If the matrices A and B commute, i.e., if $AB = BA$, then*

$$e^{A+B} = e^A e^B.$$

---

[7] It is important to note that exp(3*eye(2)) gives an incorrect result, since exp is array smart and computes the exponential of each entry of the matrix. Try it! The command expm is one command in MATLAB which is *not* array smart. Instead expm is matrix smart, and it must be used when computing the exponential of a matrix. Symbolic Toolbox users will want to try expm(3*sym(eye(2))).

For example, note that A = [1,2;-1,0] and B = [0,-2;1,1] commute (enter A*B and B*A and compare). Compute

```
>> expm(A + B)
ans =
    2.7183          0
         0     2.7183


>> expm(A)*expm(B)
ans =
    2.7183          0
         0     2.7183
```

to verify that $e^{A+B} = e^A e^B$ for these particular matrices A and B. However, matrices C = [1,1;1,1] and D = [1,2;3,4] do not commute (Check this with C*D and D*C.). Use MATLAB to show that $e^{C+D} \neq e^C e^D$.

Another property that we will find extremely useful is the following.

Proposition 2. *If A is an $n \times n$ matrix, then the derivative of e^{tA} is given as*

$$\frac{d}{dt} e^{tA} = A e^{tA}.$$

We end this section with a powerful theorem. One of the first initial value problems encountered in a course on ordinary differential equations is $x' = ax$, $x(0) = x_0$. It is easy to show (separate the variables) that the solution of this initial value problem is $x = e^{at} x_0$. One would hope that a similar result applies to the matrix equation $\mathbf{x}' = A\mathbf{x}$, $\mathbf{x}(0) = \mathbf{x}_0$.

Theorem 1. *Let $\mathbf{x}' = A\mathbf{x}$ be the matrix equation representing a system of n first order, linear, homogeneous differential equations with constant coefficients. The solution of the initial value problem, $\mathbf{x}' = A\mathbf{x}$, $\mathbf{x}(0) = \mathbf{x}_0$, is*

$$\mathbf{x}(t) = e^{At} \mathbf{x}_0.$$

Readers can easily prove this theorem by direct substitution and the application of Proposition 2. The initial condition is easily verified using the result developed in Example 8.

Symbolic Toolbox Users. Owners of the Symbolic Toolbox can readily check the validity of this last proposition for the matrix A = [1,-1;-1,1].

```
>> syms t
>> A = sym([1,-1;-1,1]);
>> diff(expm(t*A),t)
ans =
[  exp(2*t), -exp(2*t)]
[ -exp(2*t),  exp(2*t)]
>> A*expm(t*A)
ans =
[  exp(2*t), -exp(2*t)]
[ -exp(2*t),  exp(2*t)]
```

Note that the results are identical, showing that $\frac{d}{dt} e^{tA} = A e^{tA}$ for this particular matrix A. Experiment! Convince yourself that $\frac{d}{dt} e^{tA} = A e^{tA}$ for other square matrices A. Try a 3×3 matrix.

In Example 4, we used eigenvalues and eigenvectors to show that the solution of the system

$$\begin{bmatrix} x_1 \\ x_2 \\ x_3 \end{bmatrix}' = \begin{bmatrix} 8 & -5 & 10 \\ 2 & 1 & 2 \\ -4 & 4 & -6 \end{bmatrix} \begin{bmatrix} x_1 \\ x_2 \\ x_3 \end{bmatrix},$$

with initial condition $\mathbf{x}_0 = [2, 2, -3]^T$, was

$$\begin{bmatrix} x_1(t) \\ x_2(t) \\ x_3(t) \end{bmatrix} = \begin{bmatrix} 6e^{-2t} - 4e^{3t} \\ -4e^{3t} + 6e^{2t} \\ -6e^{-2t} + 3e^{2t} \end{bmatrix}.$$

Let's use Theorem 1 to produce the same result.

```
>> syms t
>> A = [8 -5 10;2 1 2;-4 4 -6];
>> x0 = sym([2;2;-3]);
>> x = expm(A*t)*x0
x =
[ 6*exp(-2*t)-4*exp(3*t)]
[ 6*exp(2*t)-4*exp(3*t)]
[ 3*exp(2*t)-6*exp(-2*t)]
```

This is a startling result, clearly demonstrating the power of the exponential matrix.

Solving Systems with Repeated Eigenvalues

We have shown how to solve the system $\mathbf{x}' = A\mathbf{x}$ when the eigenvalues of A are distinct. The matrix A in Example 6 had a repeated eigenvalue, but it still had a full complement of eigenvectors, and that allowed us to complete the solution. However, if the matrix A is $n \times n$ and there are fewer than n linearly independent eigenvectors, then the level of difficulty is raised an order or two in magnitude.

The problem. The following example is typical of 2×2 systems with only one eigenvalue.

Example 9. *Find the solution of the initial value problem*

$$\begin{aligned} x_1' &= 13x_1 + 11x_2, \\ x_2' &= -11x_1 - 9x_2, \end{aligned} \tag{12.16}$$

with initial conditions $x_1(0) = 1$ and $x_2(0) = 2$.

System (12.16) has the matrix form

$$\mathbf{x}' = \begin{bmatrix} x_1 \\ x_2 \end{bmatrix}' = \begin{bmatrix} 13 & 11 \\ -11 & -9 \end{bmatrix} \begin{bmatrix} x_1 \\ x_2 \end{bmatrix} = A\mathbf{x}, \tag{12.17}$$

with initial condition $\mathbf{x}(0) = [1, 2]^T$. It is easily verified (try poly(A)) that the characteristic polynomial is $p(\lambda) = \lambda^2 - 4\lambda + 4 = (\lambda - 2)^2$, indicating that A has an eigenvalue $\lambda = 2$ of algebraic multiplicity two. Difficulties become immediately apparent when we execute the command

```
>> null(A - 2*eye(2),'r')
ans =


    -1
     1
```

showing that, up to multiplication by a constant, $\mathbf{v}_1 = [-1, 1]^T$ is the only eigenvector. We need two independent solutions to form the general solution of system (12.16), but we only have one.

We need some terminology to describe this phenomenon.

Definition. *If λ_j is an eigenvalue of the matrix A, we define the* **algebraic multiplicity** *of λ_j to be the largest power of $(\lambda - \lambda_j)$ which divides the characteristic polynomial of A. We define the* **geometric multiplicity** *of λ_j to be the dimension of the nullspace of $A - \lambda_j I$, i.e. the dimension of the* **eigenspace** *associated to λ_j.*

It can be shown that the geometric multiplicity is never larger than the algebraic multiplicity. In the case of system (12.17), the geometric multiplicity (one) of the eigenvalue $\lambda = 2$ is strictly smaller than its algebraic multiplicity (two). Comparison of Examples 6 and 9 show that we have a problem precisely when the geometric multiplicity is smaller than the algebraic multiplicity.

Solving systems with only one eigenvalue. We will complete the solution of Example 9 by computing e^{tA} for the matrix A in (12.17). To do so, we notice that $tA = 2tI + t(A - 2I)$. Furthermore, since the identity matrix commutes with every other matrix, the matrices $2tI$ and $t(A - 2I)$ commute. Hence, we can use Proposition 1 and the identity developed in Example 8 to write

$$e^{tA} = e^{2tI + t(A-2I)} = e^{2tI} e^{t(A-2I)} = e^{2t} I e^{t(A-2I)} = e^{2t} e^{t(A-2I)}. \tag{12.18}$$

Using the definition of the exponential of a matrix in equation (12.18) to compute the second exponential, we get

$$e^{tA} = e^{2t} \left(I + t(A - 2I) + \frac{t^2}{2!}(A - 2I)^2 + \frac{t^3}{3!}(A - 2I)^3 + \cdots \right) \tag{12.19}$$

For this matrix we get a break, since $\lambda = 2$ is the *only* eigenvalue of the 2×2 matrix A. It can be shown if an $n \times n$ matrix A has only one eigenvalue λ, then $(A - \lambda I)^n$ is the zero matrix[8]. We can verify this in this case by executing

```
>> (A - 2*eye(2))^2
ans =
     0     0
     0     0
```

---

[8] This is not easy. See Exercises 54–57

Therefore, in (12.19), the powers $(A - 2I)^k = 0$ for all $k \geq 2$ and the series truncates after two terms, leaving

$$e^{tA} = e^{2t}\,(I + t(A - 2I))\,,$$

$$= e^{2t}\left\{\begin{bmatrix} 1 & 0 \\ 0 & 1 \end{bmatrix} + t\left(\begin{bmatrix} 13 & 11 \\ -11 & -9 \end{bmatrix} - 2\begin{bmatrix} 1 & 0 \\ 0 & 1 \end{bmatrix}\right)\right\}\,,$$

$$= e^{2t}\begin{bmatrix} 1 + 11t & 11t \\ -11t & 1 - 11t \end{bmatrix}.$$

According to Theorem 1, the solution of system (12.17), with initial condition $\mathbf{x}(0) = [1, 2]^T$, is given by $\mathbf{x}(t) = e^{tA}\,\mathbf{x}(0)$. Thus,

$$\mathbf{x}(t) = e^{tA}\,\mathbf{x}(0),$$

$$= e^{2t}\left(\begin{bmatrix} 1 + 11t & 11t \\ -11t & 1 - 11t \end{bmatrix}\right)\begin{bmatrix} 1 \\ 2 \end{bmatrix},$$

$$= e^{2t}\begin{bmatrix} 1 + 33t \\ 2 - 33t \end{bmatrix}.$$

Symbolic Toolbox Users. Owners of the Symbolic Toolbox can easily check this last solution.

```
>> syms t
>> A = sym([13 11;-11 -9]);
>> x0 = sym([1;2]);
>> x = expm(t*A)*x0
x =
[     exp(2*t)+33*t*exp(2*t)]
[ -33*t*exp(2*t)+2*exp(2*t)]
```

For the system (12.17) we were able to actually compute the exponential of the matrix tA. It is rare that this is so easy, but the method we used works for any 2 by 2 matrix which has a single eigenvalue λ of multiplicity 2. In such a case we always have

$$e^{tA} = e^{\lambda t}\,(I + t(A - \lambda I))\,,$$

and the solution to the initial value problem is given by

$$\mathbf{x}(t) = e^{tA}\,\mathbf{x}_0 = e^{\lambda t}\,(I + t(A - \lambda I))\,\mathbf{x}_0.$$

More generally, the method works for an $n \times n$ system as long as the matrix has only one eigenvalue of algebraic multiplicity n. In such a case, the argument leading to (12.19) yields

$$e^{tA} = e^{\lambda t}\left(I + t(A - \lambda I) + \frac{t^2}{2!}(A - \lambda I)^2 + \frac{t^3}{3!}(A - \lambda I)^3 + \cdots\right)$$

$$= e^{\lambda t}\left(I + t(A - \lambda I) + \frac{t^2}{2!}(A - \lambda I)^2 + \frac{t^3}{3!}(A - \lambda I)^3 + \cdots + \frac{t^{n-1}}{(n-1)!}(A - \lambda I)^{n-1}\right),$$

since $(A - \lambda I)^k = 0$ for $k \geq n$.

Example 10. *Find the solution of the initial value problem*

$$x_1' = 3x_1 + 3x_2 + x_3$$
$$x_2' = -7x_1 - 6x_2 - x_3 \qquad (12.20)$$
$$x_3' = -x_1 - x_2 - 3x_3,$$

with initial conditions $x_1(0) = x_2(0) = x_3(0) = 1$.

System (12.20) has matrix form

$$\mathbf{x}' = \begin{bmatrix} x_1 \\ x_2 \\ x_3 \end{bmatrix}' = \begin{bmatrix} 3 & 3 & 1 \\ -7 & -6 & -1 \\ -1 & -1 & -3 \end{bmatrix} \begin{bmatrix} x_1 \\ x_2 \\ x_3 \end{bmatrix} = A\mathbf{x},$$

with initial condition $\mathbf{x}(0) = [1, 1, 1]^T$. Matrix A has a single eigenvalue with algebraic multiplicity 3.

```
>> A=[3 3 1; -7 -6 -1; -1 -1 -3]; p=factor(poly(sym(A)))
p =
(x+2)^3
```

Hence, $(A + 2I)^3 = 0$ (check this) and the series expansion for e^{tA} truncates.

$$e^{tA} = e^{-2t}\left(I + t(A + 2I) + \frac{t^2}{2!}(A + 2I)^2\right)$$

$$= e^{-2t}\left(\begin{bmatrix} 1 & 0 & 0 \\ 0 & 1 & 0 \\ 0 & 0 & 1 \end{bmatrix} + t\begin{bmatrix} 5 & 3 & 1 \\ -7 & -4 & -1 \\ -1 & -1 & -1 \end{bmatrix} + \frac{t^2}{2!}\begin{bmatrix} 3 & 2 & 1 \\ -6 & -4 & -2 \\ 3 & 2 & 1 \end{bmatrix}\right),$$

Simplifying,

$$e^{tA} = e^{-2t}\begin{bmatrix} 1 + 5t + (3/2)t^2 & 3t + t^2 & t + (1/2)t^2 \\ -7t - 3t^2 & 1 - 4t - 2t^2 & -t - t^2 \\ -t + (3/2)t^2 & -t + t^2 & 1 - t + (1/2)t^2 \end{bmatrix}.$$

Finally, the solution to the initial value problem is found by computing

$$\mathbf{x} = e^{tA}\mathbf{x}_0 = e^{-2t}\begin{bmatrix} 1 + 5t + (3/2)t^2 & 3t + t^2 & t + (1/2)t^2 \\ -7t - 3t^2 & 1 - 4t - 2t^2 & -t - t^2 \\ -t + (3/2)t^2 & -t + t^2 & 1 - t + (1/2)t^2 \end{bmatrix}\begin{bmatrix} 1 \\ 1 \\ 1 \end{bmatrix}.$$

Thus,

$$\mathbf{x} = \begin{bmatrix} x_1 \\ x_2 \\ x_3 \end{bmatrix} = \begin{bmatrix} e^{-2t}(1 + 9t + 3t^2) \\ e^{-2t}(1 - 12t - 6t^2) \\ e^{-2t}(1 - 3t + 3t^2) \end{bmatrix}.$$

The most general case. Our first example will point the way to the general method of solution.

Example 11. *Find a fundamental set of solutions to* $\mathbf{x}' = A\mathbf{x}$, *where*

$$A = \begin{bmatrix} 1 & 0 & 1 \\ -6 & -1 & -3 \\ -6 & 1 & -5 \end{bmatrix}.$$

We will proceed as we have in previous examples. Enter A into MATLAB and compute its eigenvalues.

```
>> A=[1 0 1;-6 -1 -3; -6 1 -5]; eig(A)
ans =
    -1.0000
    -2.0000
    -2.0000
```

Notice that there are only two eigenvalues -1 and -2, and that -2 is repeated, indicating that it has multiplicity 2. Next we find eigenvectors associated with the eigenvalues

```
>> v1 = null(A-(-1)*eye(3),'r')
v1 =
    -0.5000
     1.0000
     1.0000
>> v2 = null(A-(-2)*eye(3),'r')
v2 =
    -0.3333
     1.0000
     1.0000
```

From this information we get the two solutions

$$\mathbf{x}_1(t) = e^{-t}\mathbf{v}_1 = e^{-t}\begin{bmatrix} -1/2 \\ 1 \\ 1 \end{bmatrix} \quad \text{and} \quad \mathbf{x}_2(t) = e^{-2t}\mathbf{v}_2 = e^{-2t}\begin{bmatrix} -1/3 \\ 1 \\ 1 \end{bmatrix}.$$

However, we need a third solution which is linearly independent of the first two in order to finish the job. What made our solutions in Examples 9 and 10 possible was that in each case the coefficient matrix A had only one eigenvalue λ, and satisfied $(A - \lambda I)^n = 0$, where n was the dimension of the system. From this we were able to conclude that the infinite series in (12.19) was actually a finite series which we were able to compute. We cannot expect to be so lucky this time.

In the case at hand, A has two different eigenvalues, -1 and -2 so we will not be able to find any number λ for which $(A - \lambda I)^k = 0$, for large values of k, which is necessary to make the series

$$e^{tA} = e^{\lambda t I}e^{t(A-\lambda I)} = e^{\lambda t}\left(I + t(A - \lambda I) + \frac{t^2}{2!}(A - \lambda I)^2 + \frac{t^3}{3!}(A - \lambda I)^3 + \cdots\right) \qquad (12.21)$$

reduce to a finite series. However, it is not really necessary that (12.21) truncate in this manner. Let's change our approach slightly and apply (12.21) to a vector $\mathbf{w}$.

$$e^{tA}\mathbf{w} = e^{\lambda t}\left(I\mathbf{w} + t(A - \lambda I)\mathbf{w} + \frac{t^2}{2!}(A - \lambda I)^2\mathbf{w} + \frac{t^3}{3!}(A - \lambda I)^3\mathbf{w} + \cdots\right) \qquad (12.22)$$

For this series to truncate, it is only necessary that $(A - \lambda I)^k\mathbf{w} = \mathbf{0}$ for all large k, *for the particular vector* $\mathbf{w}$.

Perhaps we can take advantage of this. Since the eigenvalue -2 has multiplicity 2, we might be able to find a vector $\mathbf{w}$ which is not a multiple of $\mathbf{v}_2$ and satisfies $(A - (-2)I)^2\mathbf{w} = 0$. We compute the nullspace of $(A - (-2)I)^2$ with

```
>>  null((A - (-2)*eye(3))^2,'r')
ans =
    -0.3333           0
     1.0000           0
          0      1.0000
```

We are in luck! The nullspace of $(A - (-2)I)^2$ has dimension 2, with the indicated vectors as a basis. Let's take $\mathbf{v}_3 = [0, 0, 1]^T$, the simplest vector in this space which is not a multiple of $\mathbf{v}_2$.[9] Using (12.21) we see that the solution with this initial value is[10]

$$\mathbf{x}_3(t) = e^{tA}\mathbf{v}_3 = e^{-2t}\ (\mathbf{v}_3 + t(A - (-2)I)\mathbf{v}_3)$$

$$= e^{-2t}\left(\begin{bmatrix} 0 \\ 0 \\ 1 \end{bmatrix} + t\begin{bmatrix} 1 \\ -3 \\ -3 \end{bmatrix}\right)$$

$$= e^{-2t}\begin{bmatrix} t \\ -3t \\ 1 - 3t \end{bmatrix}.$$

The three solutions, $\mathbf{x}_1$, $\mathbf{x}_2$, and $\mathbf{x}_3$ form a fundamental set of solutions, and we are finished.

Generalized eigenvectors and solutions. The key to finding the third solution in Example 11 was finding the vector $\mathbf{v}_3$ which satisfied $(A - (-2)I)^2\mathbf{v}_3 = 0$. Let's give such vectors a name.

Definition. *Suppose λ is an eigenvalue of the matrix A. Any vector $\mathbf{w}$ which satisfies $(A - \lambda I)^k\,\mathbf{w} = \mathbf{0}$ for some integer k is called a* **generalized eigenvector** *associated with the eigenvalue λ.*

The fact is that there are always enough generalized eigenvectors to find a basis for solutions. We will state this result as a theorem.

Theorem 2. *Suppose A is a matrix and λ is an eigenvalue of algebraic multiplicity d. Then there is an integer $k \leq d$ such that the set of all generalized eigenvectors associated to λ is equal to the nullspace of $(A - \lambda I)^k$ and has dimension equal to d, the algebraic multiplicity of λ.*

In Example 9, we found that $(A - 2I)^2 = 0$. This meant that every 2-vector was a generalized eigenvector associated to the eigenvalue 2. It was the fact that every vector was a generalized eigenvector that made things so easy in that case.

As a result of Theorem 2, we see that if λ is an eigenvalue of algebraic multiplicity d, then we can find d linearly independent solutions. Since the sum of the algebraic multiplicities is equal to the

---

[9] Any vector which is not a multiple of $\mathbf{v}_2$ would do. Taking the simplest one makes the computations easier.

[10] We use MATLAB to compute that $(A - (-2)I)\mathbf{v}_3 = [1, -3, -3]^T$.

dimension of the system, we have found a method for finding the general solution which works for all systems. Let's summarize the method.

Procedure for solving $x' = Ax$.

- Find the eigenvalues of A.
- For each eigenvalue λ:
    - Find the algebraic multiplicity d.
    - Find the smallest integer k for which the nullspace of $(A - \lambda I)^k$ has dimension d.
    - Find a basis $\mathbf{v}_1, \mathbf{v}_2, \ldots, \mathbf{v}_d$ of the nullspace of $(A - \lambda I)^k$.
    - For $j = 1, 2, \ldots, d$ set

$$\mathbf{x}_j(t) = e^{tA}\mathbf{v}_j$$
$$= e^{\lambda t}\left[\mathbf{v}_j + t(A - \lambda I)\mathbf{v}_j + \frac{t^2}{2!}(A - \lambda I)^2\mathbf{v}_j + \frac{t^3}{3!}(A - \lambda I)^3\mathbf{v}_j + \cdots \right. \tag{12.23}$$
$$\left. + \frac{t^{k-1}}{(k-1)!}(A - \lambda I)^{k-1}\mathbf{v}_j\right]$$

- If λ is complex, the real and imaginary parts of $\mathbf{x}_j$ will be real solutions.

This procedure will result in sufficiently many linearly independent solutions to generate the general solution.

We will do one final example illustrating the procedure.

Example 12. *Solve the initial value problem*

$$\begin{aligned}
x_1' &= 3x_1 - 3x_2 - 6x_3 + 5x_4, \\
x_2' &= -3x_1 + 2x_2 + 5x_3 - 4x_4, \\
x_3' &= 2x_1 - 6x_2 - 4x_3 + 7x_4, \\
x_4' &= -3x_1 + 5x_3 - 2x_4,
\end{aligned} \tag{12.24}$$

where $x_1(0) = 2$, $x_2(0) = -1$, $x_3(0) = 0$, and $x_4(0) = 0$.

Our system can be written as $\mathbf{x}' = A\mathbf{x}$, where

$$A = \begin{bmatrix} 3 & -3 & -6 & 5 \\ -3 & 2 & 5 & -4 \\ 2 & -6 & -4 & 7 \\ -3 & 0 & 5 & -2 \end{bmatrix}.$$

The first step is to use `eig` to compute the eigenvalues.

```
>> A = [3 -3 -6 5;-3 2 5 -4;2 -6 -4 7;-3 0 5 -2];
>> E = eig(A)
E =
  -1.0000 + 0.0000i
  -1.0000 - 0.0000i
  -1.0000
   2.0000
```

We suspect roundoff error and conjecture that -1 is an eigenvalue of multiplicity 3. If the Symbolic Toolbox is available, this is confirmed by the command

```
>> eig(sym(A))
ans =
[ 2]
[-1]
[-1]
[-1]
```

Thus, we see that A has two different eigenvalues, -1 and 2, with algebraic multiplicities 3 and 1, respectively[11].

Let's start with the difficult eigenvalue $\lambda_1 = -1$, which has mulitplicity 3. We will use `null` to compute nullspaces of powers of $A - \lambda_1 I = A + I$. It will be convenient to use `I = eye(4);`. Then

```
>> null(A + I,'r')
ans =
       -2
        1
       -1
        1
```

We see that although -1 is an eigenvalue of algebraic multiplicity 3, the eigenspace has dimension 1. Thus the eigenvalue -1 has geometric multiplicity 1.

The next step is to find the integer k for which the nullspace of $(A + I)^k$ has dimension 3. We do so by computing the nullspaces of increasing powers of $A + I$. First,

```
>> null((A + I)^2,'r')
ans =
        2        0
        0        1
        1        0
        0        1
```

shows that the dimension has increased to 2. However, we want the dimension to be 3, so we compute

```
>> null((A + I)^3,'r')
ans =
        1        0        0
        0        0        1
        0        1        0
        0        0        1
```

This time the dimension is 3, the same as the algebraic multiplicity, so $k = 3$ and we are in business.

---

[11] Owners of the Symbolic Toolbox can further confirm the algebraic multiplicity of the eigenvalues with `ps = poly(sym(A))`, `factor(ps)`.

The next step is to find a basis for the nullspace of $(A + I)^3$. There are many ways to do this. We could, for example, choose the three column vectors found by `null((A + I)^3,'r')`.[12] However, there is a systematic way to proceed that will reduce the algebra somewhat. Here is the plan:

- Choose a vector $\mathbf{v}_1$ which is in the nullspace of $(A+I)^3$, but **not** in the nullspace of $(A+I)^2$. That is, $\mathbf{v}_1$ must not be a linear combination of the basis vectors found for the nullspace of $(A + I)^2$. Of course, this makes $(A + I)^3\mathbf{v}_1 = \mathbf{0}$.
- Set $\mathbf{v}_2 = (A + I)\mathbf{v}_1$. Then $(A + I)^2\mathbf{v}_2 = (A + I)^3\mathbf{v}_1 = \mathbf{0}$.
- Set $\mathbf{v}_3 = (A + I)\mathbf{v}_2$. Then $(A + I)\mathbf{v}_3 = (A + I)^2\mathbf{v}_2 = \mathbf{0}$.

The result is that the vectors are automatically linearly independent,[13] and because of the algebraic relations between them the computations become a lot easier.

There are many possibilities in the choice of $\mathbf{v}_1$. When this is true, it is usually good to choose the simplest one. We choose $v_1 = [1, 0, 0, 0]^T$. In MATLAB we execute

```
>> v1 = [1 0 0 0]';
>> v2 = (A + I)*v1;
>> v3 = (A + I)*v2;
```

The vectors $\mathbf{v}_1$, $\mathbf{v}_2$, and $\mathbf{v}_3$ are a basis for the nullspace of $(A + I)^3$, and satisfy

$$\mathbf{v}_1 = \begin{bmatrix} 1 \\ 0 \\ 0 \\ 0 \end{bmatrix}, \quad \mathbf{v}_2 = (A+I)\mathbf{v}_1 = \begin{bmatrix} 4 \\ -3 \\ 2 \\ -3 \end{bmatrix}, \quad \text{and} \quad \mathbf{v}_3 = (A+I)\mathbf{v}_2 = (A+I)^2\mathbf{v}_1 = \begin{bmatrix} -2 \\ 1 \\ -1 \\ 1 \end{bmatrix}. \quad (12.25)$$

We use (12.23) to compute the solution to the initial value problem with each of these initial conditions. The first computation is the hardest, but notice how the algebraic relations of the basis vectors in (12.25) speeds things up.

$$\mathbf{x}_1(t) = e^{tA}\,\mathbf{v}_1 = e^{-t}\left(\mathbf{v}_1 + t(A + I)\,\mathbf{v}_1 + \frac{t^2}{2}(A + I)^2\mathbf{v}_1\right) = e^{-t}\left(\mathbf{v}_1 + t\mathbf{v}_2 + \frac{t^2}{2}\mathbf{v}_3\right)$$

$$= e^{-t}\left(\begin{bmatrix} 1 \\ 0 \\ 0 \\ 0 \end{bmatrix} + t\begin{bmatrix} 4 \\ -3 \\ 2 \\ -3 \end{bmatrix} + \frac{t^2}{2}\begin{bmatrix} -2 \\ 1 \\ -1 \\ 1 \end{bmatrix}\right) = e^{-t}\begin{bmatrix} 1 + 4t - t^2 \\ -3t + t^2/2 \\ 2t - t^2/2 \\ -3t + t^2/2 \end{bmatrix}.$$

The other computations are shorter. Since $(A + I)^2\mathbf{v}_2 = \mathbf{0}$, we have

$$\mathbf{x}_2(t) = e^{tA}\,\mathbf{v}_2 = e^{-t}\left(\mathbf{v}_2 + t(A + I)\,\mathbf{v}_2\right) = e^{-t}\left(\mathbf{v}_2 + t\mathbf{v}_3\right)$$

$$= e^{-t}\left(\begin{bmatrix} 4 \\ -3 \\ 2 \\ -3 \end{bmatrix} + t\begin{bmatrix} -2 \\ 1 \\ -1 \\ 1 \end{bmatrix}\right) = e^{-t}\begin{bmatrix} 4 - 2t \\ -3 + t \\ 2 - t \\ -3 + t \end{bmatrix}.$$

---

[12] We could also start by choosing an eigenvector, then take a vector from the nullspace of $(A + I)^2$ that is not a multiple of the eigenvector, and finally take a vector in the nullspace of $(A + I)^3$ which is linearly independent of the first two choices.

[13] We recommend that you prove this yourself.

Finally, since $(A + I)\mathbf{v}_3 = \mathbf{0}$, $\mathbf{v}_3$ is an eigenvector, and

$$\mathbf{x}_3(t) = e^{tA}\mathbf{v}_3 = e^{-t}\mathbf{v}_3 = e^{-t}\begin{bmatrix} -2 \\ 1 \\ -1 \\ 1 \end{bmatrix}.$$

Thus, we have found three linearly independent solutions corresponding to the eigenvalue $\lambda_1 = -1$, which has algebraic multiplicity three. We get one more in the usual way from the eigenvalue $\lambda_2 = 2$, which has algebraic multiplicity 1.

```
v4 = null(A - 2*I,'r')
v4 =
    -0.5000
     0.5000
     0.5000
     1.0000
```

The solution, $\mathbf{x}_4$, to the initial value problem with initial value $\mathbf{x}_4(0) = \mathbf{v}_4$ is

$$\mathbf{x}_4(t) = e^{tA}\mathbf{v}_4 = e^{2t}\mathbf{v}_4 = e^{2t}\begin{bmatrix} -1/2 \\ 1/2 \\ 1/2 \\ 1 \end{bmatrix}.$$

Having found four linearly independent solutions, linearity allows us to form the general solution of system (12.24) by forming a linear combination of our basic solutions.

$$\mathbf{x}(t) = c_1\mathbf{x}_1(t) + c_2\mathbf{x}_2(t) + c_3\mathbf{x}_3(t) + c_4\mathbf{x}_4(t) \tag{12.26}$$

To solve the initial value problem with $\mathbf{x}(0) = \mathbf{x}_0 = [2, -1, 0, 0]^T$, we must substitute $t = 0$ in equation (12.26).

$$\mathbf{x}(0) = c_1\mathbf{x}_1(0) + c_2\mathbf{x}_2(0) + c_3\mathbf{x}_3(0) + c_4\mathbf{x}_4(0)$$

$$\mathbf{x}_0 = c_1\mathbf{v}_1 + c_2\mathbf{v}_2 + c_3\mathbf{v}_3 + c_4\mathbf{v}_4$$

Of course, this is equivalent to the matrix equation

$$\mathbf{x}_0 = [\mathbf{v}_1, \mathbf{v}_2, \mathbf{v}_3, \mathbf{v}_4]\begin{bmatrix} c_1 \\ c_2 \\ c_3 \\ c_4 \end{bmatrix},$$

so we can find the solution with the following MATLAB code

```
>> V = [v1 v2 v3 v4];
>> x0 = [2;-1;0;0];
>> c = V\x0
c =
     5.0000
     3.0000
     7.0000
     2.0000
```

Therefore, the solution of system (12.20) is $\mathbf{x}(t) = 5\mathbf{x}_1(t) + 3\mathbf{x}_2(t) + 7\mathbf{x}_3(t) + 2\mathbf{x}_4(t)$, or

$$\mathbf{x}(t) = 5e^{-t}\begin{bmatrix} 1 + 4t - t^2 \\ -3t + t^2/2 \\ 2t - t^2/2 \\ -3t + t^2/2 \end{bmatrix} + 3e^{-t}\begin{bmatrix} 4 - 2t \\ -3 + t \\ 2 - t \\ -3 + t \end{bmatrix} + 7e^{-t}\begin{bmatrix} -2 \\ 1 \\ -1 \\ 1 \end{bmatrix} + 2e^{2t}\begin{bmatrix} -1/2 \\ 1/2 \\ 1/2 \\ 1 \end{bmatrix}.$$

Users of the Symbolic Toolbox might want to compare this last solution with

```
>> syms t
>> A = sym(A);
>> x0 = sym(x0);
>> x = expm(A*t)*x0
x =
[       3*exp(-t)+14*t*exp(-t)-5*t^2*exp(-t)-exp(2*t)]
[ -12*t*exp(-t)+5/2*t^2*exp(-t)+exp(2*t)-2*exp(-t)]
[     7*t*exp(-t)-5/2*t^2*exp(-t)-exp(-t)+exp(2*t)]
[ -12*t*exp(-t)+5/2*t^2*exp(-t)-2*exp(-t)+2*exp(2*t)]
```

Exercises

For the matrices in Exercises 1 – 4, perform each of the following tasks.

   i Use the command p = poly(A) to find the characteristic polynomial.

  ii Plot the characteristic polynomial p with the commands

```
>> t = linspace(-4,4);
>> y = polyval(p,t);
>> plot(t,y)
>> grid on
>> axis([-4,4,-10,10])
```

and find the roots of the polynomial with r = roots(p).

  iii Find the eigenvalues with e = eig(A) and compare the eigenvalues with the roots of the polynomial found in part (ii). Where are the eigenvalues located on the graph produced in part (ii)?

1. $\begin{bmatrix} -8 & 10 \\ -5 & 7 \end{bmatrix}$ 2. $\begin{bmatrix} -4 & 6 & 0 \\ -3 & 5 & 0 \\ -2 & 0 & 1 \end{bmatrix}$ 3. $\begin{bmatrix} -5 & -18 & 12 \\ -3 & -8 & 6 \\ -6 & -18 & 13 \end{bmatrix}$ 4. $\begin{bmatrix} 2 & -3 & -1 & -3 \\ 3 & -4 & 3 & -3 \\ 0 & 0 & 3 & 0 \\ -3 & 3 & -3 & 2 \end{bmatrix}$

Symbolic Toolbox Users. In Exercises 5 – 8, enter the matrix from the indicated exercise as a symbolic object, as in A = sym([-8,10;-5,7]). Find the characteristic polynomial with p = poly(A), plot with ezplot(p), set the window and grid with axis([-4,4,-10 10]), grid on, factor with factor(p), and find the roots with solve(p). Finally, find the eigenvalues with eig(A) and compare with the roots of the characteristic polynomial.

  5. Exercise 1   6. Exercise 2   7. Exercise 3   8. Exercise 4

  9. Given that $\lambda = -1$ is an eigenvalue of the matrix

$$A = \begin{bmatrix} 3 & 6 & 10 \\ 0 & -1 & 0 \\ 0 & -1 & -2 \end{bmatrix},$$

reduce the augmented matrix M = [A - (-1)*eye(3),zeros(3,1)] and interpret the result to find the associated eigenvector. Use null(A - (-1)*eye(3),'r') and compare results. Perform similar analysis for the remaining eigenvalues, $\lambda = -2$ and $\lambda = 3$.

10. Use [V,E] = eig(A) to find the eigenvalues and eigenvectors of the matrix A in Exercise 9. Strip off column three of matrix V with v3 = V(:,3). Set v3 = v3/v3(3). In a similar manner, show that the eigenvectors in the remaining columns of matrix V are multiples of the eigenvectors found in Exercise 9.

11. **Symbolic Toolbox Users.** Use [V,E] = eig(sym(A)) to find the eigenvalues and eigenvectors of the matrix A from Exercise 9. Compare with the results found in Exercise 9.

12. Given that $\lambda = 1$ is an eigenvalue of matrix

$$A = \begin{bmatrix} 5 & 5 & -12 & 18 \\ 4 & 6 & -12 & 18 \\ 0 & 6 & -11 & 12 \\ -2 & 2 & -3 & 1 \end{bmatrix},$$

reduce the augmented matrix M = [A - 1*eye(4),zeros(4,1)] and interpret the result to find a basis for the associated eigenspace. Use null(A - 1*eye(4),'r') and compare the results. Perform similar analysis for the remaining eigenvalue, $\lambda = -2$. Clearly state the algebraic and geometric multiplicity of each eigenvalue.

For the matrices in Exercises 13 – 19, find the eigenvalues and eigenvectors. You may use any method you wish as long as you choose eigenvectors with all integer entries. For each eigenvalue state the algebraic and the geometric multiplicity.

13. $\begin{bmatrix} -6 & 0 \\ 7 & 1 \end{bmatrix}$
14. $\begin{bmatrix} 0 & 1 \\ 1 & 0 \end{bmatrix}$
15. $\begin{bmatrix} 4 & 1 \\ -4 & 0 \end{bmatrix}$
16. $\begin{bmatrix} 2 & -1 \\ 1 & 0 \end{bmatrix}$

17. $\begin{bmatrix} -2 & 2 & -1 \\ -6 & 5 & -2 \\ 4 & -2 & 3 \end{bmatrix}$
18. $\begin{bmatrix} -3 & -2 & 3 \\ -2 & 1 & 1 \\ -7 & -3 & 6 \end{bmatrix}$
19. $\begin{bmatrix} -1 & 4 & -10 & 2 \\ 2 & -1 & 7 & -1 \\ 2 & -3 & 9 & -1 \\ 0 & -2 & 2 & 3 \end{bmatrix}$

Real, Distinct Eigenvalues. If the eigenvalues of the matrix A are real and distinct, then the eigenvectors will be independent, providing the appropriate number of independent solutions for $\mathbf{x}' = A\mathbf{x}$ (see Example 4). Find the general solution of each of the systems in Exercises 20 – 22. Please choose eigenvectors with integer entries.

20. $\begin{aligned} x' &= -10x + 7y \\ y' &= -14x + 11y \end{aligned}$

21. $\begin{aligned} x' &= -15x + 18y - 8z \\ y' &= -4x + 4y - 2z \\ z' &= 16x - 21y + 9z \end{aligned}$

22. $\begin{aligned} x_1' &= 12x_1 - 16x_2 - 10x_3 - 22x_4 \\ x_2' &= 2x_1 - 4x_2 - 2x_3 - 4x_4 \\ x_3' &= -5x_1 + 9x_2 + 7x_3 + 10x_4 \\ x_4' &= 8x_1 - 11x_2 - 8x_3 - 15x_4 \end{aligned}$

Symbolic Toolbox Users. In Exercises 23 – 25, check the solution of the system from the indicated exercise with the dsolve command (see Chapter 10). For example, the solution of the system in Exercise 20 is obtained with the command [x,y] = dsolve('Dx = -10*x+7*y','Dy = -14*x+11*y','t').

23. Exercise 20.
24. Exercise 21.
25. Exercise 22.

Algebraic Multiplicity Equals Geometric Multiplicity. If the algebraic multiplicity of each eigenvalue equals its geometric multiplicity you will have sufficient eigenvectors to find the general solution (see Example 6). In Exercises 26 – 28, find the general solution of the indicated system. Please choose eigenvectors with integer entries.

26. $\begin{aligned} x' &= -5x - 3y - 3z \\ y' &= -6x - 2y - 3z \\ z' &= 12x + 6y + 7z \end{aligned}$

27. $\begin{aligned} w' &= 16w - 12x + 3y + 3z \\ x' &= 36w - 26x + 6y + 6z \\ y' &= 18w - 12x + y + 3z \\ z' &= 36w - 24x + 6y + 4z \end{aligned}$

28. $\begin{aligned} w' &= w + 3x + 3z \\ x' &= 3w + x + 3z \\ y' &= -6w + y - 6z \\ z' &= -3w - 3x - 5z \end{aligned}$

Complex Eigenvalues. The systems in Exercises 29 – 31 have eigenvalues that are complex (see Example 7). Find the general solution. Please choose eigenvectors whose real and imaginary parts contain integer entries.

208

29. $x' = -7x + 15y$
    $y' = -6x + 11y$

30. $x' = 4x + 2y + z$
    $y' = -10x - 5y - 2z$
    $z' = 10x + 6y + 2z$

31. $w' = w + x - 2y - z$
    $x' = -5w - 2x + 6y + 3z$
    $y' = -5w - 2x + 5y + 2z$
    $z' = 10w + 5x - 10y - 4z$

Initial Value Problems. In Exercises 32–34, you are given a matrix A and initial condition x_0. Find a solution to the initial value problem $\mathbf{x}' = A\mathbf{x}$, $\mathbf{x}(0) = \mathbf{x}_0$. Eliminate fractions and decimals by using eigenvectors whose real and imaginary parts contain integer entries.

32. $\begin{bmatrix} 2 & 5 & -5 \\ -3 & -4 & 3 \\ -3 & -1 & 0 \end{bmatrix}, \begin{bmatrix} 1 \\ 1 \\ 1 \end{bmatrix}$

33. $\begin{bmatrix} -31 & 42 & 14 \\ -14 & 18 & 7 \\ -28 & 42 & 11 \end{bmatrix}, \begin{bmatrix} 1 \\ 2 \\ 1 \end{bmatrix}$

34. $\begin{bmatrix} -6 & 2 & 1 & 3 \\ -10 & 5 & 2 & 4 \\ 0 & 0 & 1 & 0 \\ -11 & 2 & 1 & 6 \end{bmatrix}, \begin{bmatrix} 3 \\ 2 \\ 1 \\ 1 \end{bmatrix}$

Symbolic Toolbox Users. In Exercises 35–37, use the `dsolve` command to solve the initial value problem from the indicated exercise and compare with the solution found there. *Note:* It is definitely challenging to get these solutions to match. Be persistent!

35. Exercise 32.

36. Exercise 33.

37. Exercise 34.

Numerical Versus Exact. In Exercises 38–40, plot all components of the solution to the system found in the indicated exercise against time on the same graph over the interval $[0, 2\pi]$. Then find a numerical solution over the same time interval $[0, 2\pi]$ using `ode45`. Plot all components of the numerical solution to the system against time on a second graph. How well do the exact and numerical solutions compare?

38. Exercise 32.

39. Exercise 33.

40. Exercise 34.

In Exercises 41–43, use definition (12.15) of the exponential matrix to compute the indicated matrix. If the Symbolic Toolbox is available, use it to check your results.

41. e^A, where $A = \begin{bmatrix} 0 & 0 \\ 0 & 0 \end{bmatrix}$

42. e^{At}, where $A = \begin{bmatrix} 0 & 5 \\ 0 & 0 \end{bmatrix}$

43. e^{At}, where $A = \begin{bmatrix} 2 & 0 \\ 0 & 3 \end{bmatrix}$

44. Create a diagonal matrix D with the command D = diag([1,2,3]). It is easy to calculate powers of a diagonal matrix. Enter D^2, D^3, D^4, etc. Do you see a pattern emerging? If

$$D = \begin{bmatrix} \lambda_1 & 0 & \cdots & 0 \\ 0 & \lambda_2 & \cdots & 0 \\ \vdots & \vdots & \ddots & \vdots \\ 0 & 0 & \cdots & \lambda_n \end{bmatrix},$$

then what is D^k?

If a matrix A has distinct eigenvalues (no repeated eigenvalues), then the matrix A can be *diagonalized*; i.e., there exists a diagonal matrix D and an invertible matrix V such that $A = VDV^{-1}$. Furthermore, if $A = VDV^{-1}$ where D is a diagonal matrix, then D contains the eigenvalues of the matrix A on its diagonal and the columns of matrix V contain eigenvectors of A associated with the eigenvalues in the corresponding columns of matrix D. For the matrices A in Exercises 45–47, compute the eigenvalues and eigenvectors with [V,D] = eig(A), then show that $A = VDV^{-1}$ by comparing A with V*D*inv(V).

45. $A = \begin{bmatrix} 0 & 2 & -3 \\ 1 & 1 & 1 \\ 2 & -2 & 5 \end{bmatrix}$

46. $A = \begin{bmatrix} -5 & -6 & -3 \\ 14 & 11 & 7 \\ -10 & -2 & -4 \end{bmatrix}$

47. $A = \begin{bmatrix} -7 & 4 & 8 & -2 \\ -4 & 5 & 6 & -2 \\ -10 & 6 & 14 & -5 \\ -20 & 16 & 32 & -13 \end{bmatrix}$

Show that if $A = VDV^{-1}$ where D is a diagonal marix, then $A^k = VD^kV^{-1}$ for every positive integer k. In Exercises 45–47, use this fact to compute A^5 for the matrix A in the indicated exercise. Use MATLAB to compare A^5 with V*D^5*inv(V).

48. Exercise 45.

49. Exercise 46.

50. Exercise 47.

Show that if $A = VDV^{-1}$ where D is a diagonal marix, then $e^A = Ve^D V^{-1}$. (*Hint:* Use use the previous three exercises and equation (12.15)). Further, show that $e^{tA} = Ve^{tD}V^{-1}$ and use this result to calculate e^{tA} for the matrices in Exercises 51 – 53. If the Symbolic Toolbox available, check the result with `expm(t*A)`.

51. $A = \begin{bmatrix} -3 & 4 \\ -2 & 3 \end{bmatrix}$　　　52. $A = \begin{bmatrix} -8 & 9 \\ -6 & 7 \end{bmatrix}$　　　53. $A = \begin{bmatrix} 8 & 6 \\ -12 & -10 \end{bmatrix}$

The Cayley-Hamilton theorem states that every matrix must satisfy its characteristic polynomial p_A. For example, the characteristic polynomial of

$$A = \begin{bmatrix} 5 & -5 & 3 \\ 2 & -1 & 2 \\ 0 & 1 & 2 \end{bmatrix}$$

is $p_A(\lambda) = \lambda^3 - 6\lambda^2 + 11\lambda - 6$. The following commands will verify that $p_A(A) = A^3 - 6A^2 + 11A - 6I = 0$.

```
>> A = [5 -5 3;2 -1 2;0 1 2];
>> A^3 - 6*A^2 + 11*A - 6*eye(3)
```

In Exercises 54 – 57, verify the Cayley-Hamilton theorem for the given matrix. Note: You may also want to investigate the `polyvalm` command.

54. $\begin{bmatrix} 5 & 2 \\ -6 & -2 \end{bmatrix}$　　55. $\begin{bmatrix} -1 & 1 \\ 6 & 4 \end{bmatrix}$　　56. $\begin{bmatrix} -4 & 1 & 1 \\ 11 & 0 & -6 \\ -12 & 2 & 4 \end{bmatrix}$　　57. $\begin{bmatrix} -1 & 1 & 0 \\ -1 & -2 & -1 \\ 3 & 2 & 2 \end{bmatrix}$

If A is 2×2 and has an eigenvalue c of multiplicity two, then the characteristic polynomial is $p_A(\lambda) = (\lambda - c)^2$. Because $p_A(A) = 0$ (see the Cayley-Hamilton theorem), $(A - cI)^2 = 0$. Each of the 2×2 matrices in Exercises 58 – 61 has an eigenvalue c of multiplicity two. Find that eigenvalue c and verify that $(A - cI)^2 = 0$.

58. $\begin{bmatrix} 1 & 1 \\ -1 & 3 \end{bmatrix}$　　59. $\begin{bmatrix} -3 & -1 \\ 1 & -1 \end{bmatrix}$　　60. $\begin{bmatrix} 1 & -4 \\ 1 & 5 \end{bmatrix}$　　61. $\begin{bmatrix} -6 & 1 \\ -1 & -4 \end{bmatrix}$

Repeated Eigenvalue. Use MATLAB to show that the matrices in Exercises 62 – 65 have an eigenvalue of algebraic multiplicity two, but geometric multiplicity one. Use the identity $tA = \lambda tI + t(A - \lambda I)$, where λ is the eigenvalue of the matrix A, and the technique of Example 9 to compute the exponential matrix e^{tA}.

62. $A = \begin{bmatrix} -4 & 1 \\ -1 & -2 \end{bmatrix}$　　63. $A = \begin{bmatrix} -6 & -1 \\ 1 & -4 \end{bmatrix}$　　64. $A = \begin{bmatrix} 3 & -1 \\ 1 & 1 \end{bmatrix}$　　65. $A = \begin{bmatrix} 6 & -4 \\ 1 & 2 \end{bmatrix}$

Symbolic Toolbox Users. In Exercises 66 – 69, use the Symbolic Toolbox commands `syms t, expm(t*sym(A))` to find e^{tA} for the matrix in the indicated exercise.

66. Exercise 62.　　　67. Exercise 63.　　　68. Exercise 64.　　　69. Exercise 65.

Repeated Eigenvalue. Use MATLAB to show that the systems in Exercises 70 and 71 have a repeated eigenvalue of algebraic multiplicity two, but geometric multiplicity one. Compute the exponential matrix. Use the exponential matrix and the technique of Example 9 to compute the solution of the given initial value problem.

70. $\mathbf{x}' = \begin{bmatrix} -3 & 1 \\ -4 & 1 \end{bmatrix} \mathbf{x}, \quad \mathbf{x}_0 = \begin{bmatrix} 1 \\ 1 \end{bmatrix}$　　　71. $\mathbf{x}' = \begin{bmatrix} 7 & -1 \\ 1 & 5 \end{bmatrix} \mathbf{x}, \quad \mathbf{x}_0 = \begin{bmatrix} -1 \\ 2 \end{bmatrix}$

Symbolic Toolbox Users. In Exercises 72 and 73, compute the solution to the initial value problem in the indicated exercise using the use the Symbolic Toolbox and the exponential matrix. *Note:* It is definitely challenging to get these solutions to match. Be persistent!

72. Exercise 70.　　　　　73. Exercise 71.

74. The matrix

$$A = \begin{bmatrix} 3 & -1 & -2 \\ 14 & -5 & -4 \\ 11 & -2 & -7 \end{bmatrix}$$

has a repeated eigenvalue of algebraic multiplicity three. Use the technique of Example 9 to compute e^{tA}. Hint: See Exercises 20–21. Use the result to solve the initial value problem

$$x_1' = 3x_1 - x_2 - 2x_3,$$

$$x_2' = 14x_1 - 5x_2 - 4x_3,$$

$$x_3' = 11x_1 - 2x_2 - 7x_3,$$

where $x_1(0) = x_2(0) = x_3(0) = 1$.

The initial value problems in Exercises 75 – 77 have a coefficient matrix with a repeated eigenvalue. Use the technique of Examples 11 and 12 to find the solution. If the Symbolic Toolbox is available, you can check your answer with the `dsolve` command.

75. $x' = 6x - 8y + 7z$
$y' = 3x - 4y + 3z$
$z' = -x + 2y - 2z$
$x(0) = y(0) = z(0) = 1$

76. $x' = x + y - 3z$
$y' = 10x - 2y - 10z$
$z' = -2x + y$
$x(0) = y(0) = z(0) = 1$

77. $x_1' = 3x_1 + 4x_2 - 3x_3 - 12x_4$
$x_2' = 2x_1 - 4x_2 + 6x_3 + 15x_4$
$x_3' = -4x_1 - 5x_2 + 4x_3 + 15x_4$
$x_4' = 2x_1 + x_2 - 2x_4$
$x_1(0) = x_2(0) = x_3(0) = x_4(0) = 1$

In Exercises 78 – 88 find a fundamental set of solutions to $\mathbf{x}' = A\mathbf{x}$. Solve the initial value problem with $\mathbf{x}(0) = \mathbf{x}_0$.

78. $A = \begin{bmatrix} -5 & -3 \\ 6 & 1 \end{bmatrix}$, $\mathbf{x}_0 = \begin{bmatrix} 2 \\ 3 \end{bmatrix}$

79. $A = \begin{bmatrix} -3 & 0 & -1 \\ 4 & -7 & 4 \\ 8 & -10 & 6 \end{bmatrix}$, $\mathbf{x}_0 = \begin{bmatrix} 0 \\ 2 \\ 4 \end{bmatrix}$

80. $A = \begin{bmatrix} -3 & 1 & 0 \\ -5 & 1 & 0 \\ -9 & 3 & -2 \end{bmatrix}$, $\mathbf{x}_0 = \begin{bmatrix} 1 \\ 1 \\ -2 \end{bmatrix}$

81. $A = \begin{bmatrix} -2 & -10 & 8 \\ -1 & -1 & -2 \\ -3 & -5 & -2 \end{bmatrix}$, $\mathbf{x}_0 = \begin{bmatrix} 15 \\ -4 \\ 0 \end{bmatrix}$

82. $A = \begin{bmatrix} 1 & 4 & -4 \\ 0 & -1 & 0 \\ 1 & 2 & -3 \end{bmatrix}$, $\mathbf{x}_0 = \begin{bmatrix} 1 \\ 0 \\ 0 \end{bmatrix}$

83. $A = \begin{bmatrix} 2 & 2 & 1 \\ -7 & -5 & -3 \\ 5 & 2 & 2 \end{bmatrix}$, $\mathbf{x}_0 = \begin{bmatrix} -2 \\ 3 \\ 0 \end{bmatrix}$

84. $A = \begin{bmatrix} -8 & -7 & 10 & -2 \\ 7 & 6 & -12 & 2 \\ 1 & 1 & -3 & 0 \\ 5 & 3 & -2 & -1 \end{bmatrix}$, $\mathbf{x}_0 = \begin{bmatrix} -2 \\ -8 \\ -6 \\ 1 \end{bmatrix}$

85. $A = \begin{bmatrix} -6 & 0 & 7 & 3 \\ -11 & -5 & 21 & 9 \\ -8 & -3 & 14 & 7 \\ 7 & 5 & -16 & -9 \end{bmatrix}$, $\mathbf{x}_0 = \begin{bmatrix} 2 \\ -1 \\ -1 \\ 3 \end{bmatrix}$

86. $A = \begin{bmatrix} 17 & -9 & 14 & 51 \\ 12 & -6 & 9 & 34 \\ 16 & -11 & 14 & 50 \\ -8 & 5 & -7 & -25 \end{bmatrix}$, $\mathbf{x}_0 = \begin{bmatrix} -2 \\ -2 \\ 1 \\ 0 \end{bmatrix}$

87. $A = \begin{bmatrix} -9 & 7 & -27 & 17 \\ 7 & -4 & 18 & -13 \\ 4 & -1 & 8 & -7 \\ -1 & 4 & -9 & 3 \end{bmatrix}$, $\mathbf{x}_0 = \begin{bmatrix} 5 \\ 0 \\ 0 \\ 2 \end{bmatrix}$

88. $A = \begin{bmatrix} 2 & -4 & -1 & 6 & 3 \\ -53 & 47 & 12 & -89 & -33 \\ -13 & 12 & 2 & -22 & -9 \\ -39 & 36 & 9 & -67 & -25 \\ 18 & -16 & -4 & 30 & 10 \end{bmatrix}$, $\mathbf{x}_0 = \begin{bmatrix} -5 \\ 0 \\ 0 \\ 3 \\ 0 \end{bmatrix}$

13.   Advanced Use of PPLANE6

We introduced `pplane6` in Chapter 7. In this chapter, we will investigate some of the more advanced features of `pplane6`, such as drawing nullclines, finding and classifying equilibrium points, displaying linearizations, and drawing the separatrices that accompany saddle points.

Recall that `pplane6` can provide numerical solutions of planar, autonomous systems of the form

$$x' = f(x, y), \tag{13.1}$$
$$y' = g(x, y). \tag{13.2}$$

For the purpose of discussion in this chapter, we will assume three things:

- the vertical axis is the y-axis,
- the horizontal axis is the x-axis, and
- the independent variable is t.

We learned in Chapter 7 that `pplane6` allows you to change these defaults, but we will not do so here.

Nullclines

Suppose that we take the right-hand side of equation (13.1) and set it equal to zero. The curve defined by $f(x, y) = 0$ can be drawn in the xy-phase plane. Along this curve the vector field has the form $[f(x, y), g(x, y)]^T = [0, g(x, y)]^T$. The 0 in the first component means that the vector field must point either directly up or directly down, depending on the sign of $g(x, y)$. The solution curve through this point is tangent to the vector field, and therefore must be moving either up or down. This reflects the fact that $x' = f(x, y) = 0$ along the curve. The curve defined by $f(x, y) = 0$ is called the *x-nullcline*.

It might be useful to look at the solution curve $y \rightarrow [x(t), y(t)]^T$ as the path of a particle. The velocity vector for this particle is $[x', y']^T = [f(x, y), g(x, y)]^T$. Therefore, the points on the x-nullcline, defined by $f(x, y) = 0$, are precisely those points where x', the component of the velocity in the x direction, is equal to zero. When this is the case the motion of the particle must be vertical.

For similar reasons, along the curve defined by setting the right-hand side of equation (13.2) equal to zero, i.e., $g(x, y) = 0$, the vector field must point either directly to the left or to the right. This curve is called the *y-nullcline*. The solution curve through any point on the y-nullcline must be moving horizontally.

Confused? Perhaps an example will clear the waters.

Example 1. *Plot the nullclines for the system*

$$\frac{dx}{dt} = 2x + 3y - 6,$$
$$\frac{dy}{dt} = 3x - 4y - 12. \tag{13.3}$$

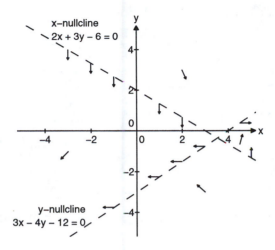

Figure 13.1. The nullclines of $dx/dt = 2x + 3y - 6$ and $dy/dt = 3x - 4y - 12$.

The x-nullcline is defined by the equation $2x + 3y - 6 = 0$, and is pictured in Figure 13.1. The y-nullcline, defined by $3x - 4y - 12 = 0$, is also shown in Figure 13.1. Notice that at the origin, $(x, y) = (0, 0)$, $2x + 3y - 6 = -6 < 0$. Hence, we must have $dx/dt = 2x + 3y - 6 < 0$ everywhere below the x-nullcline, and $dx/dt = 2x + 3y - 6 > 0$ everywhere above the x-nullcline. In particular, these inequalities are valid along the y-nullcline. These considerations are reflected in the right or left direction of the vector field arrows plotted on the y-nullcline in Figure 13.1. Similar considerations allow us to determine the up or down direction of the vector field along the x-nullcline.

More information is available from Figure 13.1. We can make a rough estimate of the direction of the vector field in each of the four areas bounded by the nullclines. For example, in the area above both nullclines, the vector field must point down and to the right (southeast), because the vector field varies continuously, and therefore in any region bounded by the nullclines the direction must interpolate between the directions on the nullcline boundaries.

Pplane6 completely automates the process of finding nullclines and the direction of the vector field. Begin by entering the equations $x' = 2x + 3y - 6$ and $y' = 3x - 4y - 12$ in the PPLANE6 Setup window. Set the display window so that $-5 \leq x \leq 5$ and $-5 \leq y \leq 5$, as shown in Figure 13.2.

Look carefully at Figure 13.2, and note that we have selected the **Nullclines** option. Click the **Proceed** button to transfer information to the PPLANE6 Display window. Use your mouse and click the phase plane in several locations to start solution trajectories. You might not get exactly the same image shown in Figure 13.3, but you should get an image that demonstrates three very important points.

- The nullclines are drawn in Figure 13.3, and arrows have been drawn on the nullclines indicating the direction that trajectories will follow when passing through the nullclines. There are fewer arrows than in Figure 13.1, but they point in the same direction as shown there.
- Solution trajectories move up or down along the x-nullcline, and move left or right along the y-nullcline.
- The nullclines are plotted in dashes. This is to emphasize that nullclines are not solution curves, which are plotted as solid curves.

213

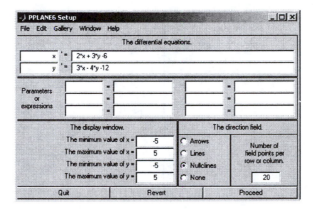

Figure 13.2. Setting up $x' = 2x + 3y - 6$ and $y' = 3x - 4y - 12$.

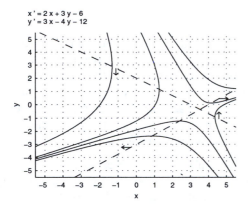

Figure 13.3. The nullclines in the PPLANE6 Display window

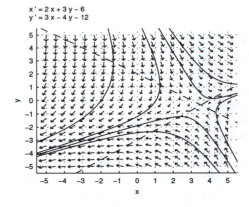

Figure 13.4. A second view of the null-clines.

There is another way to display nullclines using pplane6. Return to the Setup window, select the **Arrows** option, and then click **Proceed**. This will remove the nullclines from the Display window and add the vector field. Next select **Solutions→Show nullclines**, in the Display window and the nullclines will be plotted on top of the vector field as shown in Figure 13.4. You might find this view more revealing, if more cluttered.

Equilibrium Points

Consider again the planar, autonomous system

$$x' = f(x, y),$$
$$y' = g(x, y).$$

214

A point (x, y) in the phase plane is called an *equilibrium point* if both right-hand sides vanish, i.e., if $f(x, y) = 0$ and $g(x, y) = 0$. This means that equilibrium points are the points of intersection of the nullclines.

Example 2. *Find all equilibrium points of the system in Example 1.*

Setting both of the right-hand sides in (13.3) equal to zero yields the system

$$2x + 3y = 6,$$
$$3x - 4y = 12.$$

This linear system is easily solved, either by hand or using MATLAB, to find that $x = 60/17 \approx 3.5294$ and $y = -6/17 \approx -0.3529$.

Pplane6 can also be used to find equilibrium points. Re-enter the system of Example 1 in the PPLANE6 Setup window as shown in Figure 13.2. Select the **Nullclines** for the direction field, click **Proceed**, then start a few trajectories by clicking the mouse in the phase plane. Select **Solutions→Find an equilibrium point**, then move the mouse "cross hairs" near the intersection of the nullclines and click the mouse. Pplane6 will plot the equilibrium point, as shown in Figure 13.5.

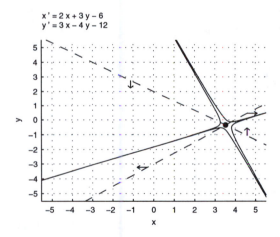

Figure 13.5. The equilibrium point in Example 2 is the intersection point of the nullclines.

Pplane6 also opens the PPLANE6 Equilibrium point data window, which summarizes a number of important facts about the equilibrium point. The first entry says "There is a saddle point at (3.5294, −0.35294)." The location agrees with what we found earlier. The equilibrium point is called a *saddle point* because, if you have a topographic map of an area including a mountain pass or a saddle, the topographic lines of equal altitude look something like the solution curves in Figure 13.5. Click the **Go Away** button to close the window.

Example 3. *Find all equilibrium points of the system*

$$x' = x^2 + y^2 - 4,$$
$$y' = x - y^2. \tag{13.4}$$

Setting the right-hand sides of the differential equations in (13.4) equal to zero, we obtain the system

$$x^2 + y^2 - 4 = 0,$$
$$x - y^2 = 0.$$

This system is nonlinear, but it can be solved by substituting $y^2 = x$ in the first equation, and solving the resulting quadratic equation, finding $x = (-1 \pm \sqrt{17})/2$. For each solution there are two possible values of y satisfying $y^2 = x$. Since one of the values of x is negative, the only viable solutions are

$$\left(\frac{-1 + \sqrt{17}}{2}, \frac{\sqrt{-2 + 2\sqrt{17}}}{2} \right) \approx (1.5616, 1.2496), \quad \text{and}$$

$$\left(\frac{-1 + \sqrt{17}}{2}, -\frac{\sqrt{-2 + 2\sqrt{17}}}{2} \right) \approx (1.5616, -1.2496).$$

(13.5)

Symbolic Toolbox Users. Owners of the Symbolic Toolbox can use the `solve` command to find the equilibrium points.

```
>> [x,y]=solve('x^2+y^2-4=0','x-y^2=0','x,y')
x =
[ -1/2+1/2*17^(1/2)]
[ -1/2+1/2*17^(1/2)]
[ -1/2-1/2*17^(1/2)]
[ -1/2-1/2*17^(1/2)]
y =
[   1/2*(-2+2*17^(1/2))^(1/2)]
[ -1/2*(-2+2*17^(1/2))^(1/2)]
[   1/2*(-2-2*17^(1/2))^(1/2)]
[ -1/2*(-2-2*17^(1/2))^(1/2)]
```

The first two entries of vectors x and y contain symbolic results identical to our algebraic results in (13.5). To find the numerical values, change the class of x and y to `double`.

```
>> S=[double(x),double(y)]
S =
    1.5616                1.2496
    1.5616               -1.2496
   -2.5616            0 + 1.6005i
   -2.5616            0 - 1.6005i
```

The first two rows of S contain the equilibrium points; the last two rows of S are complex, and are ignored.

Of course, `pplane6` should provide identical results. Enter the system in (13.4) into the PPLANE6 Setup window. Set the display window so that $-5 \le x \le 5$ and $-5 \le y \le 5$. Select **Nullclines** for the direction field, then click **Proceed** to transfer information to the PPLANE6 Display window. Notice that the x-nullcline is a circle, and the y-nullcline is a parabola. They have two points of intersection. These are the equilibrium points and they can be found using **Solutions→Find an equilibrium point**. Click the mouse in the phase plane to start several solution trajectories near each equilibrium point, similar to the image shown in Figure 13.6.

Notice that solutions near the upper of the two equilibrium points behave like those near the saddle point in Example 2, while the solution curves near the lower equilibrium point behave quite differently. Select **Solutions→Find an equilibrium point** and click the mouse near the upper equilibrium point. Repeat the process with the lower equilibrium point. This will open the PPLANE6 Equilibrium point data window shown in Figure 13.7. This window indicates that the lower equilibrium point is located at approximately $(1.5616, -1.2496)$, in agreement with (13.5). Since the solution trajectories in Figure 13.6 spiral away from the lower equilibrium point, it is called a *spiral source*.

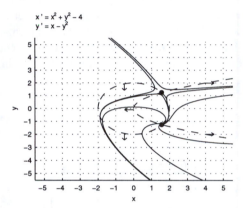

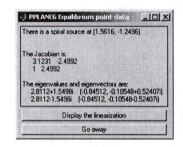

Figure 13.6. The two equilibrium points in Example 3 are a saddle point and a source.

Figure 13.7. Description of the lower equilibrium point.

Linear Systems

The system

$$x' = ax + by,$$
$$y' = cx + dy, \tag{13.6}$$

is planar, autonomous, and linear. The system can be written in the form

$$\mathbf{x}' = \begin{bmatrix} x \\ y \end{bmatrix}' = \begin{bmatrix} a & b \\ c & d \end{bmatrix} \begin{bmatrix} x \\ y \end{bmatrix} = A\mathbf{x}.$$

The equilibrium points are found by setting the right-hand sides of both differential equations in (13.6) equal to zero, leading to the system

$$\begin{matrix} ax + by = 0, \\ cx + dy = 0, \end{matrix} \quad \text{or} \quad A\mathbf{x} = \mathbf{0}. \tag{13.7}$$

Thus, the equilibrium points are the vectors in the nullspace of the coefficient matrix A. If $\det A \neq 0$, the system in (13.7) has a unique solution, so the origin $(x, y) = (0, 0)$ is the only equilibrium point of the system in (13.6). On the other hand, if det $A = 0$, then the system has many equilibrium points. For example, the system $x' = x + 2y$, $y' = 2x + 4y$ has an entire line of equilibrium points. Experiment with this system in `pplane6`.

Example 4. *Find and classify the equilibrium point of the linear system*

$$x' = -2x - 2y,$$
$$y' = 3x + y. \tag{13.8}$$

Select **Gallery→linear system** from the PPLANE6 Setup menu. Notice that most of the work has been done for you. All you have to do is adjust the parameters A, B, C, and D. Compare system (13.8) with the general linear system (13.6) and note that $a = -2$, $b = -2$, $c = 3$, and $d = 1$. Enter these parameters for A, B, C, and D in the PPLANE6 Setup window, then click **Proceed** to transfer information to the PPLANE6 Display window. Select **Solutions→Keyboard input** and start a solution trajectory with initial condition $(0, 1)$. Note how this solution trajectory "spirals inward" toward the origin, as shown in Figure 13.8.

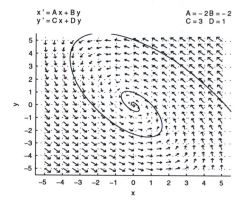
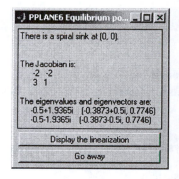

Figure 13.8. A spiral sink at the equilibrium point $(0, 0)$.

Figure 13.9. Description of the equilibrium point in Figure 13.8.

Select **Solutions→Find an equilibrium point** and click your mouse near the origin in the phase plane. Let's carefully examine the information provided in the PPLANE6 Equilibrium point data window shown in Figure 13.9.

- The first entry provides the type of the equilibrium point and the location, in this case a spiral sink at $(0,0)$. This certainly agrees with our view of the trajectory in Figure 13.8.
- Next we are presented with the the Jacobian matrix,

$$J = \begin{bmatrix} -2 & -2 \\ 3 & 1 \end{bmatrix}$$

218

in this case. If system (13.8) is written in matrix form $\mathbf{x}' = A\mathbf{x}$, then the Jacobian J is identical to the coefficient matrix A. This is always the case for linear systems.

- The eigenvalue-eigenvector pairs of the Jacobian are reported. In this case the eigenvalues are a complex conjugate pair and the eigenvectors are also complex.

- Finally, there is a **Display the linearization** button. This will provide no new information in this case, since system in (13.8) is already linear.

Nonlinear Systems

Any planar, autonomous system that is not in the precise form of (13.6) is nonlinear. Nonlinear systems are much harder to analyze than linear systems since they rarely possess analytical solutions. Therefore, we must use qualitative analysis and numerical solutions. There are many techniques for analyzing nonlinear systems qualitatively. One of the most useful is the analysis of the behavior of solutions near equilibrium points using the *linearization*.

If the term linearization sounds familiar, it should. In your first calculus class, you constantly analyzed the behavior of a function at a point by approximating the function with a first order Taylor's series, whose graph is the line through the point on the graph with the same slope as the graph. This line is called the *tangent line*, and its slope is equal to the derivative of the function. The graph of the tangent line closely approximates that of the function near the point of tangency, but as you move away from the point of tangency, the approximation is less accurate. Therefore the technique of linearization only provides local information about the behavior of the function.

We will use the same technique to analyze a nonlinear system near an equilibrium point. We will use the first order Taylor series approximation of the right-hand side of the system. The analog of the derivative is called the Jacobian. For some systems, the linearization will enable us to predict the behavior of solutions near the equilibrium point. Again, as we move away from the equilibrium point, the approximation deteriorates, so the information is only of value locally, by which we mean near the equilibrium point.

Definition. *Suppose that a planar vector field is defined by*

$$\mathbf{F}(x, y) = \begin{bmatrix} f(x, y) \\ g(x, y) \end{bmatrix}.$$

Then the Jacobian of $\mathbf{F}$ *is defined to be*

$$\mathbf{JF}(x, y) = \begin{bmatrix} f_x(x, y) & f_y(x, y) \\ g_x(x, y) & g_y(x, y) \end{bmatrix},$$

where $f_x(x, y)$ *denotes the partial derivative of* f *with respect to* x, *evaluated at* (x, y), *etc.*

Example 5. *Consider the nonlinear system of differential equations*

$$\begin{aligned} x' &= 4x - 2x^2 - xy, \\ y' &= y - y^2 + 2xy. \end{aligned} \tag{13.9}$$

Find the Jacobian of the vector field.

The right-hand side of the system in (13.9) can be written as the vector field

$$\mathbf{F}(x, y) = \begin{bmatrix} f(x, y) \\ g(x, y) \end{bmatrix} = \begin{bmatrix} 4x - 2x^2 - xy \\ y - y^2 + 2xy \end{bmatrix}.$$

The Jacobian matrix of $\mathbf{F}$ is

$$\mathbf{JF}(x, y) = \begin{bmatrix} f_x(x, y) & f_y(x, y) \\ g_x(x, y) & g_y(x, y) \end{bmatrix} = \begin{bmatrix} 4 - 4x - y & -x \\ 2y & 1 - 2y + 2x \end{bmatrix}. \tag{13.10}$$

Symbolic Toolbox Users. The Symbolic Toolbox can compute the Jacobian.

```
>> syms x y
>> F=[4*x-2*x^2-x*y;y-y^2+2*x*y]
F =
[ 4*x-2*x^2-x*y]
[   y-y^2+2*x*y]
>> JF=jacobian(F)
JF =
[   4-4*x-y,            -x]
[       2*y,  1-2*y+2*x]
```

You can also use the command subs to evaluate the Jacobian at a particular point.

```
>> subs(JF,{x,y},{3/4,5/2})
ans =
   -1.5000    -0.7500
    5.0000    -2.5000
```

Consequently,

$$\mathbf{JF}(3/4, 5/2) = \begin{bmatrix} -1.5 & -0.75 \\ 5 & -2.5 \end{bmatrix}. \tag{13.11}$$

Pplane6 can greatly simplify the analysis of nonlinear systems.

Example 6. *The system*

$$\begin{aligned} x' &= 2x(2 - x) - xy, \\ y' &= y(1 - y) + 2xy. \end{aligned} \tag{13.12}$$

models the interaction of two populations, a rabbit population x and a fox population y. Find and analyze the equilibrium points of this system.

A little algebra reveals that the system in (13.12) is identical to the system in Example 5. We present the system in this new form to facilitate discussion of the model. If we deleted the xy terms from the right-hand sides in (13.12), both the rabbit and fox populations would grow according to a logistic model. The xy terms measures the effect of interaction between the two species. The presence of the term $-xy$ in the first equation indicates that interaction between the species is harmful to the rabbit population, while the presence of the term $+2xy$ in the second equation indicates that the interaction increases the growth rate of the fox population. Because the interaction among the species is beneficial to one but harmful to the other, this type of system is sometimes called a *predator-prey* model.

Enter the system in (13.12) into the PPLANE6 Setup window. Set the display window so that $0 \le x \le 3$ and $0 \le y \le 3$. Select the **Arrows** radio button for the direction field and click **Proceed** to transfer information to the PPLANE6 Display window.

In the PPLANE6 Display window, select **Solutions**→**Show nullclines** to overlay the nullclines on the vector field. Use **Solutions**→**Find an equilibrium point** to locate all four equilibrium points of the rabbit-fox system. The results are shown in Figure 13.10.

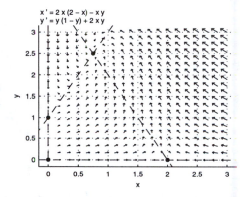

Figure 13.10. Phase plane and equilibrium points for the rabbit-fox populations.

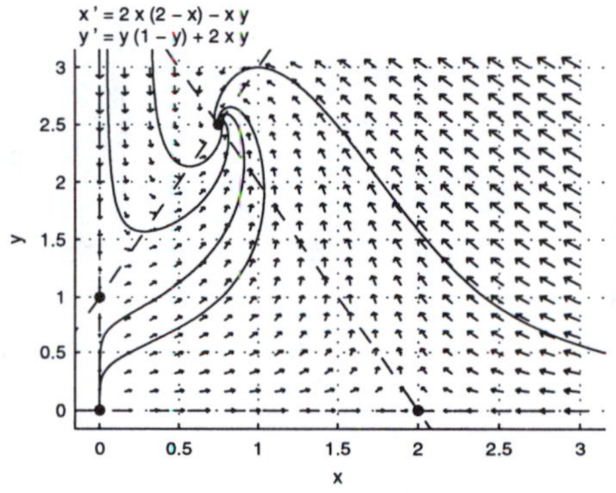

Figure 13.11. Phase portrait for the rabbit-fox system.

Once you've found all of the equilibrium points, you can obtain a summary of the equilibrium points you found by selecting **Solutions**→**List computed equilibrium points**. A summary of the computed equilibrium points will be printed in the MATLAB command window[1].

```
(0.7500, 2.5000)    Spiral sink.
(0.0000, 1.0000)    Saddle point.
(0.0000, 0.0000)    Nodal source.
(2.0000, 0.0000)    Saddle point.
```

Solution trajectories starting near $(0, 0)$ approach it as $t \to -\infty$, and become tangent to the y-axis

---

[1] Symbolic Toolbox owners will want to try
`[x,y]=solve('2*x*(2-x)-x*y','y*(1-y)+2*x*y','x,y')`.

in the process. An equilibrium point with the property that all solutions starting nearby tend towards it as $t \to -\infty$, and become tangent to a fixed direction is called a "nodal source."

The points $(0, 1)$ and $(2, 0)$ are saddle points. Click your mouse near these points and note that solutions tend toward these points, only to veer away at the last moment.

Finally, the equilibrium point at $(0.75, 2.5)$ is called a spiral sink. Think of the drain in a bathtub, with water swirling around and into the drain. Click your mouse near the point $(0.75, 2.5)$ to witness this behavior.

If you've carefully followed the directions in the preceding paragraphs, Figure 13.11 shows a possible phase portrait for the rabbit-fox system.

Linearization. Let revisit the spiral sink at $(0.75, 2.5)$. Select **Solutions→Find an equilibrium point** and click near the point $(0.75, 2.5)$ to re-open the PPLANE6 Equilibrium point data window, shown in Figure 13.12.

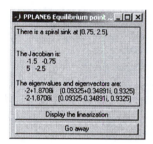

Figure 13.12. Data for the spiral sink at $(0.75, 2.5)$.

The first line in the PPLANE6 Equilibrium point data window indicates that there is a spiral sink at $(0.75, 2.5)$. The Jacobian matrix evaluated at the equilibrium point $(0.75, 2.5)$, and its eigenvalue-eigenvector pairs are displayed. Click the **Display the linearization** button to show the linear approximation of the rabbit-fox system shown in Figure 13.13.

The title heading in Figure 13.13 holds the key to the relationship between the Jacobian and the linearization. Notice that the system in Figure 13.13 is the general linear system

$$u' = Au + Bv$$
$$v' = Cu + Dv$$

The coefficients are given to the right of the system. Notice that the values of A, B, C, and D are precisely the entries in the Jacobian matrix reported in the PPLANE6 Equilibrium point data window shown in Figure 13.12. So, you've now got the secret! If you want a linearization of a nonlinear system at an equilibrium point, compute the Jacobian at the equilibrium point, and form a linear system using the entries of the Jacobian matrix.

Actually, we are hiding a little mathematics here. Note that the linearization in Figure 13.13 has an equilibrium point at $(0, 0)$, while the equilibrium point for the nonlinear system in (13.11) is at $(0.75, 2.5)$. When forming the linearization, we have to translate coordinate systems so that the origin

222

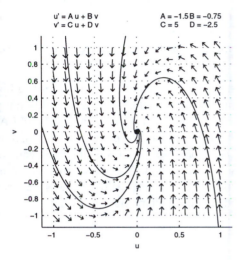

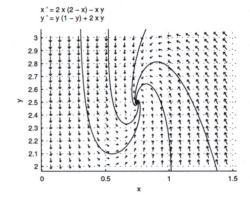

Figure 13.13. The linearization of the system at $(0.75, 2.5)$.

Figure 13.14. A closer look at the equilibrium point at $(0.75, 2.5)$.

of the linearization aligns itself with the equilibrium point $(0.75, 2.5)$. That's why we're using u and v coordinates to define the linearization. They are the coordinates of the translated coordinate system.

Zoom in square. If we compare the solutions of the linearization in Figure 13.13 to those of the nonlinear system in Figure 13.11, we see that there are many differences, although the solutions to the nonlinear system near the equilibrium point at $(0.75, 2.5)$ do resemble those of the linearization near the origin. If we remember that the linearization provides a good approximation only near the equilibrium point, we can understand that.

If we want to compare the nonlinear system and its linearization, we should zoom in near the equilibrium point for the nonlinear system. Notice that the linearization window in Figure 13.13 is square, and that both variables range from -1 to 1. In other words, the scales are the same on each axis. When you zoom in in the usual way using **Edit→Zoom in** in the PPLANE6 Display window, it is highly unlikely that the scales will be the same on the two axes. However, if you choose **Edit→Zoom in square**, you automatically get a new window with the same scale on the two axes. After making the menu selection, click the (left) mouse button on the equilibrium point is the Display window and drag a small rectangle.[2]

One possible result of this process for the system in Figure 13.11 is shown in Figure 13.14. Notice that the solution curves for the linearization in Figure 13.13 and those for the nonlinear system in Figure 13.14 are very similar. These two figures also clearly display the fact that the equilibrium points for the nonlinear system and its linearization are different. Obtaining the similarity we observe between Figures 13.13 and 13.14 may require several zooms. For some systems the attempt to achieve similarity is doomed to failure. (See Exercise 20.)

---

[2] If you have a three button mouse, you do not need to use the menu. Simply click and drag with the center button.

Classification of equilibrium points. It turns out that there is a small list of all possible equilibrium points for planar, linear systems with constant coefficients. We have seen saddle points, nodal sources, nodal sinks, spiral sources, and spiral sinks. Another is a *center*, which is an equilibrium point surrounded by closed solution curves.[3] The behavior of the solutions is completely determined by the eigenvalues and eigenvectors of the coefficient matrix.

For nonlinear systems there are more possibilities, but in many cases the type of an equilibrium point is the same as that of the linearization at that point. We see this clearly for the equilibrium point studied in Figures 13.13 and 13.14. When you find an equilibrium point using pplane6, the PPLANE6 Equilibrium point window gives as much information about the equilibrium point as it can, based on its analysis of the linearization. In some cases it gives all the information you need. In others you will be able to find more precise information by further analysis of the system. We will offer some examples in the exercises. In the Appendix to this chapter you will find a complete list of the messages that can appear in the PPLANE6 Equilibrium point window, together with an interpretation of what each message means.

Separatrices

There are four special solution curves associated with any saddle point. There are two curves that approach the saddle point at $t \to \infty$, called *stable solution curves*, and two that approach the saddle point at $t \to -\infty$, called *unstable solution curves*. Collectively, these curves are called *separatrices*.[4] This is best demonstrated with an example.

Example 7. *Consider the nonlinear system*

$$x' = x(2 - x) - 2xy,$$
$$y' = y(2 - y) - 2xy. \tag{13.13}$$

Assume that x and y represent populations that are always non-negative.

Without the xy terms, the equation for each population would be the logistic equation. However, because of the minus signs, including the xy terms in this model is detrimental to the growth of both populations. This is an example of *competition*. Think about sheep and cattle populations competing for the same grazing land.

Enter the system into the PPLANE6 Setup window, set the display window to $0 \le x \le 3$ and $0 \le y \le 3$, select **Arrows** for the direction field, and click **Proceed** to transfer information to the PPLANE6 Display window.

In the PPLANE6 Display window, select **Solutions→Show nullclines** to overlay the nullclines on the direction field. Use **Solutions→Find an equilibrium point** to find the four equilibrium points. Selecting **Solutions→List computed equilibrium points** will show the following summary in MATLAB's command window.

```
(0.0000, 2.0000)    Sink.
(0.0000, 0.0000)    Source.
(2.0000, 0.0000)    Sink.
(0.6667, 0.6667)    Saddle point.
```

---

[3] There are a number of others that appear less frequently.
[4] The singular is separatrix.

Compute some solution trajectories by clicking various initial conditions in the phase plane. You will soon discover that some of the solutions tend to the sink at $(0, 2)$, spelling extinction for species x, while other trajectories tend to the sink at $(2, 0)$, indicating extinction of species y.

Select **Solutions→Plot stable and unstable orbits** and click near the saddle point at $(0.6667, 0.6667)$. Pplane6 will plot the separatrices in green. Notice that the stable solution curves "separate" solution trajectories that tend to the sink at $(0, 2)$ from those that tend to the sink at $(2, 0)$, as shown in Figure 13.15.

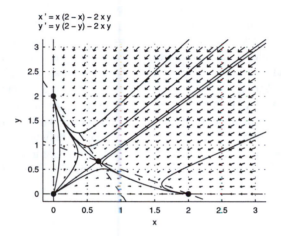

$$x' = x\,(2 - x) - 2\,x\,y$$
$$y' = y\,(2 - y) - 2\,x\,y$$

Figure 13.15. The separatrices divide the phase plane into basins of attraction.

For any sink there is a *basin of attraction*, consisting of the initial points of solutions which tend to the sink as $t \to \infty$. We see in Figure 13.15 that the stable trajectories form the boundary of the basins of attraction of the two sinks. Separatrices often separate the phase plane into areas where the solution curves behave differently. This explains the special importance of separatrices as well as their name.

Limit Sets of Solution Curves

Perhaps you have noticed that you do not have to provide a solution interval for a solution computed by pplane6, although it is necessary to do so when solving using ode45 and any other solver. Perhaps you have wondered how pplane6 knows when to stop computing. You may also have noticed the messages that appear in the message window, telling you that a solution ... left the computation window, or ...--> a possible eq. pt. near .... These messages actually reveal how pplane6 knows when to stop. It turns out that there is a very small list of the possible "endings" of a solution curve.

What you see in the PPLANE6 Display window is only a fraction of what's being computed. There is a computation window that's larger than what is actually displayed and allows solutions to leave the Display window and return. One possible ending of a computed solution is that it escapes this computation window. If it does, pplane6 stops computing. The interesting story is about what can happen to a solution that stays in the computation window forever.

What we referred to as the "ending" of a solution curve is officially called the *limit set* of the curve.

At the beginning of the twentieth century, the mathematicians Henri Poincaré and Ivar Bendixson proved that there are only three possible limit sets for a solution curve that stays in a bounded region:

- An equilibrium point.
- A closed solution curve.
- A directed planar graph with vertices that are equilibrium points and edges that are solution curves.

These are called the *Poincaré-Bendixson alternatives*. We have seen many examples of the first alternative. In each case this result was reported in the message window.

A closed solution curve can occur as the limit set in two ways. First of all, the solution curve itself can be closed. Examples of this can be found with any center. In particular, choose **vibrating spring** from the **Gallery** menu, and leave the damping term $d = 0$. Plot a few solutions. See Figure 13.16. The other possibility is that the solution curve is not closed, but approaches a closed curve that is called a *limit cycle*. Choose **van der Pol's equation** from the **Gallery** menu for an example, and plot a few solutions. See Figure 13.17. A closed solution curve repeats itself, while a solution curve which spirals toward a limit cycle, gets closer to itself in the process. pplane6 has a way of discovering when a solution curve gets very close to itself. It stops the computation when this occurs and reports in the message window that the solution curve ... --> a nearly closed orbit. The word "nearly" is there to allow for the possibility that pplane6 is making a mistake, but our experience has been that this is a rare event. Most of the time this message means that the solution curve is either closed itself or is approaching a limit cycle.

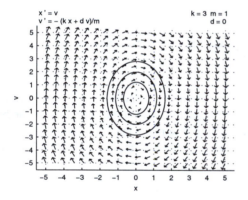

Figure 13.16. Closed orbits around a center.

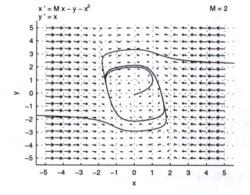

Figure 13.17. Orbits approaching a limit cycle.

If you want to plot only the "nearly closed curve," select **Solutions→Find a nearly closed orbit**, and select the direction along the solution curve in which to search. The result is reported in the message window. If such a curve is found, it is plotted in the display window and its period is reported in the message window. This feature is especially useful if you want to see what a limit cycle actually looks like. Approximately three periods of the closed curve will be plotted. See Figure 13.18.

An example of the third Poincaré-Bendixson alternative can be seen by selecting **square limit set** from the gallery. If you start a solution anywhere inside the unit square, the solution will spiral out to the boundary of the unit square, so this boundary is the limit set. Verify that the vertices of the unit square are saddle points, and the edges are separatrices. Since separatrices are solution curves, the limit set is an example of the third alternative. See Figure 13.19. In this case a solution curve starting inside the unit square is periodically getting close to the equilibrium points, and is getting closer to itself in the process. Sooner or later one of the two previously described stopping criteria is triggered, and the result is reported in the message window. Notice that when a limit set is a directed graph, pplane6 will report that the solution is getting close to an equilibrium point or to itself. Pplane6 is not yet smart enough to recognize a directed graph.

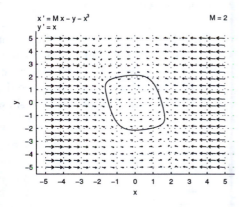

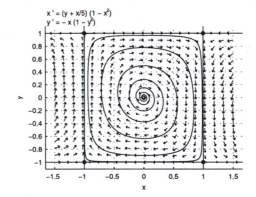

Figure 13.18. The limit cycle for the van der Pol equation.

Figure 13.19. A directed planar graph as the limit set of a solution curve.

Thus we see that the classification of Poincaré and Bendixson allows us to rely on these two stopping criteria, and do away entirely with the need to specify a solution interval. Theory does come in handy.

Other Options in PPLANE6

Solvers. If you choose **Option→Solver**, you will see a list of the ODE solvers available in pplane6. The first two are built into pplane6, and the rest is a list of all of the solvers available in MATLAB. Any of these can be used.

The Dormand-Prince solver is the default. Actually this is almost identical to ode45. However, because it is built into pplane6, Dormand-Prince can take advantage of the special nature of the program, and it is faster to access. As a result it is considerably faster than ode45. However, most modern computers are so fast that the difference is hardly noticeable.

Solver settings. When you select a new solver, or if you select **Option→Settings**, the PPLANE6 Settings window will appear. Here you can adjust the number of plot steps per computation step. This option is identical to the "Refine" option available for ode45. It sets the number of points that the solver will accurately interpolate between the computed points.

You can also adjust the relative error tolerance. Again, we will refer readers to the "Improving Accuracy" section of Chapter 8. The description of relative error tolerance in Chapter 8 describes exactly how this option performs in pplane6.[5]

---

[5] In the case of the Runge-Kutta 4 solver there is no refine option, and no tolerance. Instead you can change the step size.

In the second part of the PPLANE6 Settings window, there is a slider that will enable you to choose the speed with which solutions are computed and plotted. This is meant to be used to slow down the solution to allow you to better understand how the solution proceeds. Choose less than 10 steps per second for a full appreciation of this option.

Finally, there is a setting for determining the size of the calculation window. The default setting of 3.5 is sufficient for ordinary use of `pplane6`, but if you notice that some solutions which leave the display window do not return although you know they should, increase this parameter.

Conservative and dissipative systems and level curves. The selection **Solutions→Plot level curves** will open the PPLANE6 Level sets window, which will enable you to plot the level curves of a function in the Display window. As an example, select **pendulum** from the gallery. The total energy for the pendulum is (proportional to) $E(\theta, \omega) = \omega^2/2 - \cos\theta$. If you leave $D = 0$, **Proceed** to the display window, and select **Solutions→Plot level curves**, you will be presented with the window shown in Figure 13.16.

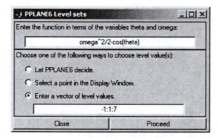

Figure 13.20. Entering the energy for the pendulum.

Notice that the first line says the function should be entered in terms of omega and `theta`, not in terms of the TEX notation \omega and \theta used in the setup window. Enter the rest of the data shown in Figure 13.16, and click **Proceed**. The sets of constant energy will be plotted. Now compute a few solutions and notice that the solution curves are identical to the level curves, demonstrating that energy is conserved in the motion of the undamped pendulum.

Return to the setup window, increase the damping constant to $D = 0.5$, and repeat the process. This time the solution curves consistently move from higher to lower energy levels, showing that damping dissipates the energy of the system.

Marking initial points. Sometimes it is advantageous to mark the initial points of the computed solutions. Choosing **Option→Mark initial points** will enable you to do just that.

Exporting solution data. If you have computed a solution curve that you want to analyze using techniques not available in `pplane6`, you can export the data to the MATLAB command window. Select **Option→ Export solution data**. If there is only one solution curve, its data will be exported. If there are more than one, you will be given the opportunity to choose which one to export with the mouse. The action is reported in the message window with a message like, `The data has been exported as the structure ppdata4 with fields t, x, and y`. A structure is a MATLAB class that is convenient for storing data. In this case the fields are denoted by `ppdata4.t`, `ppdata4.x`, and `ppdata4.y`. These are column vectors containing all of the computed points for the indicated variable. You can use the fields just like any other MATLAB variable. For example to plot the x component versus t, execute

```
>> figure, plot(ppdata4.t,ppdata4.x)
```

Editing the PPLANE6 Display window. The MATLAB editing features are somewhat hidden in the PPLANE6 Display window, because they do not mix well with the interactive features of the window. The Figure Toolbar can be activated in the **View** menu, making all of the editing features available.

Exercises

Nullclines and Equilibrium Points. In Exercises 1 – 4, without the aid of technology, using only your algebra skills, sketch the nullclines and find the equilibrium point(s) of the assigned system. Indicate the flow of the the vector field along each nullcline, similar to that shown in Figure 13.1. Check your result with `pplane6`. If the Symbolic Toolbox is available, use the `solve` command to find the equilibrium point(s).

1. $x' = x + 2y$
   $y' = 2x - y$

2. $x' = x + 2y - 4$
   $y' = 2x - y - 2$

3. $x' = 2x - y + 3$
   $y' = y - x^2$

4. $x' = 2x + y$
   $y' = 4x + 2y$

Linear systems. Exercises 5–17 are designed to help classify all potential behavior of planar, autonomous, linear systems. It is essential that you master this material before attempting the analysis of non-linear systems. Be sure to turn **off** nullclines in these exercises.

Straight line solutions. If λ, $\mathbf{v}$ is an eigenvalue-eigenvector pair of matrix A, then we know that $\mathbf{x}(t) = e^{\lambda t}\mathbf{v}$ is a solution. Moreover, if λ and $\mathbf{v}$ are real, then this solution is a line in the phase plane. Because solutions of linear systems cannot cross in the phase plane, straight line solutions prevent any sort of rotation of solution trajectories about the equilibrium point at the origin.

In Exercises 5 – 7, find the eigenvalues and eigenvectors with the `eig` and `null` commands, as demonstrated in Example 4 of Chapter 12. You may find `format rat` helpful. Then enter the system into `pplane6`, and draw the straight line solutions. For example, if one eigenvector happens to be $\mathbf{v} = [1, -2]^T$, use the Keyboard input window to start straight line solutions at $(1, -2)$ and $(-1, 2)$. Perform a similar task for the other eigenvector. Finally, the straight line solutions in these exercises divide the phase plane into four regions. Use your mouse to start several solution trajectories in each region.

5. $x' = 5x - y$
   $y' = -x + 5y$

6. $x' = 6x - y$
   $y' = -3y$

7. $x' = -x$
   $y' = 3x - 3y$

8. What condition on λ will ensure that the straight line solution $\mathbf{x}(t) = e^{\lambda t}\mathbf{v}$ moves toward the equilibrium point at the origin as time increases? What condition ensures that the straight line solution will move away from the equilibrium point as time increases?

9. If the eigenvalues and eigenvectors of a planar, autonomous, linear system are complex, then there are no straight line solutions. Use `[v,e]=eig(A)` to demonstrate that the eigenvalues and eigenvectors of the system $x' = 2.9x + 2.0y$, $y' = -5.0x - 3.1y$ are complex, then use `pplane6` to show that solution trajectories spiral in the phase plane.

10. The system $x' = x + y$, $y' = -x + 3y$ is a degenerate case, having only one straight line solution. Find the eigenvalues and eigenvectors of the system $x' = x + y$, $y' = -x + 3y$, then use `pplane6` to depict the single straight line solution. Sketch several solution trajectories and note that they almost spiral, but the single straight line solution prevents full rotation about the equilibrium point at the origin. This degenerate case is a critical case. Its trace-determinant pair lies on a curve in the trace-determinant plane that separates systems that spiral from those that do not.

Sources, sinks, and saddles. If a system has two distinct, real eigenvalues, then it will have two independent straight line solutions, and all solutions can be written as a linear combination of these two straight line solutions, as in $\mathbf{x}(t) = c_1 e^{\lambda_1 t} \mathbf{v}_1 + c_2 e^{\lambda_2 t} \mathbf{v}_2$.

11. **Nodal sink.** If a system has two distinct negative eigenvalues, then both straight line solutions will decay to the origin with the passage of time. Consequently, all solutions will decay to the origin. Enter the system, $x' = -4x + y$, $y' = -2x - y$, in pplane6 and plot the straight line solutions. Plot several more solutions and note that they also decay to the origin. Select **Solutions→Find an equilibrium point**, find the equilibrium point at the origin, then read its classification from the PPLANE6 Equilibrium point data window.

12. **Nodal source.** If a system has two distinct positive eigenvalues, then both straight line solutions will move away from the origin with the passage of time. Consequently, all solutions will move away from the origin. Enter the system, $x' = 5x - y$, $y' = 2x + 2y$, in pplane6 and plot the straight line solutions. Plot several more solutions and note that they also move away from the origin. Select **Solutions→Find an equilibrium point**, find the equilibrium point at the origin, then read its classification from the PPLANE6 Equilibrium point data window.

13. **Saddle.** If a system has one negative and one positive eigenvalue, then one straight line solution moves toward the origin and the other moves away. Consequently, general solutions (being linear combinations of straight line solutions) must do the same thing. Enter the system $x' = 9x - 14y$, $y' = 7x - 12y$, in pplane6 and plot the straight line solutions. Plot several more solutions and note that they move toward the origin only to move away at the last moment. Select **Solutions→Find an equilibrium point**, find the equilibrium point at the origin, then read its classification from the PPLANE6 Equilibrium point data window.

Spiral sources, spiral sinks, and centers. If the eigenvalues are complex, then a planar, autonomous, linear system has no straight line solutions. Consequently, trajectories are free to spiral.

14. **Spiral source.** As we saw in equation (13.12), if the real part of a complex eigenvalue is negative, the equilibrium point will be a spiral sink. What happens if the real part of the complex eigenvalue is positive? Enter the system $x' = 2.1x + 5.0y$, $y' = -1.0x - 1.9y$ in pplane6 and note that solutions spiral outward with the passage of time. Select **Solutions→Find an equilibrium point**, find the equilibrium point at the origin, note the real part of the complex eigenvalue, then note the classification of the equilibrium point in the PPLANE6 Equilibrium point data window. If your computer is too fast to differentiate an outward spiral from an inward spiral, you might try **Options→Solution direction→Forward**.

15. **Center.** If the real part of the complex eigenvalue is zero, there can be no growth or decay of solution trajectories. Enter the system $x' = -x + y$, $y' = -2x + y$, and note that pplane6 reports a nearly closed orbit in its message window. Note that pplane6 has difficulty with this type of equilibrium point, commonly called a *center*. It should report that the trajectories are purely periodic. This happens because this system is a critical case, separating linear systems with a spiral sink from systems with a spiral source. The slightest numerical error, even an error of 1×10^{-16}, makes the real part of the complex eigenvalue non-zero, indicating a spiral source or sink.

16. **Summarizing your findings.** In a table use your results from Exercises 11–15 to list conditions on the eigenvalue that predict when the equilibrium point at the origin is a nodal source, nodal sink, saddle, spiral source, spiral sink, and center.

17. **The trace-determinant plane.** It is a nice exercise to classify linear systems based on their position in the trace-determinant plane. Consider the matrix

$$A = \begin{bmatrix} a & b \\ c & d \end{bmatrix}.$$

a) Show that the characteristic polynomial of the matrix A is $p(\lambda) = \lambda^2 - T\lambda + D$, where $T = a + d$ is the trace of A and $D = \det(A) = ad - bc$ is the determinant of A.

b) We know that the characteristic polynomial factors as $p(\lambda) = (\lambda - \lambda_1)(\lambda - \lambda_2)$, where λ_1 and λ_2 are the eigenvalues. Use this and the result of part (a) to show that the product of the eigenvalues is equal to the determinant of matrix A. *Note:* This is a useful fact. For example, if the determinant is negative, then you must have one positive and one negative eigenvalue, indicating a saddle equilibrium point. Also, show that the sum of the eigenvalues equals the trace of matrix A.

230

c) Show that the eigenvalues of matrix A are given by the formula

$$\lambda = \frac{T \pm \sqrt{T^2 - 4D}}{2}.$$

Note that there are three possible scenarios. If $T^2 - 4D < 0$, then there are two complex eigenvalues. If $T^2 - 4D > 0$, there are two real eigenvalues. Finally, if $T^2 - 4D = 0$, then there is one repeated eigenvalue of algebraic multiplicity two.

d) Draw a pair of axes on a piece of poster board. Label the vertical axis D and the horizontal axis T. Sketch the graph of $T^2 - 4D = 0$ on your poster board. The axes and the parabola defined by $T^2 - 4D = 0$ divide the trace-determinant plane into six distinct regions, as shown in Figure 13.17.

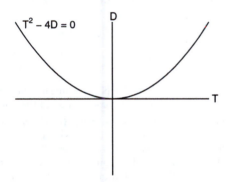

Figure 13.21. The trace-determinant plane.

e) You can classify any matrix A by its location in the trace-determinant plane. For example, if

$$A = \begin{bmatrix} 1 & 2 \\ -3 & 2 \end{bmatrix},$$

then $T = 3$ and $D = 8$, so the point (T, D) is located in the first quadrant. Furthermore, $(3)^2 - 4(8) < 0$, placing the point $(3, 8)$ above the parabola $T^2 - 4D = 0$. Finally, if you substitute $T = 3$ and $D = 8$ into the formula $\lambda = (T \pm \sqrt{T^2 - 4D})/2$, then you get eigenvalues that are complex with a positive real part, making the equilibrium point of the system $\mathbf{x}' = A\mathbf{x}$ a spiral source. Use pplane6 to generate a phase portrait of this particular system and attach the printout to the poster board at the point $(3, 8)$.

f) Linear systems possess a small number of distinctive phase portraits. Each of these is graphically different from the others, but each corresponds to the pair of eigenvalues and their multiplicities. For each case, use pplane6 to construct a phase portrait, and attach a printout at its appropriate point (T, D) in your poster board trace-determinant plane. *Hint:* There are degenerate cases on the axes and the parabola. For example, you can find degenerate cases on the parabola in the first quadrant that separate nodal sources from spiral sources. There are also a number of interesting degenerate cases at the origin of the trace-determinant plane. One final note: We have intentionally used the words "small number of distinctive cases" so as to spur argument amongst our readers when working on this activity. What do you think is the correct number?

Nonlinear Systems. We analyze the behavior of a nonlinear system near an equilibrium point with a linear approximation.

18. **The Jacobian.** Consider the nonlinear system

$$x' = x(1 - x) - xy,$$
$$y' = y(2 - y) + 2xy.$$

Show that $(-1/3, 4/3)$ is an equilibrium point of the system.

a) Without the use of technology, calculate the Jacobian of the system at the equilibrium point $(-1/3, 4/3)$. What is the equation of the linearization at this equilibrium point? Use `[v,e]=eig(J)` to find the eigenvalues and eigenvectors of this Jacobian.

b) Enter the system in `pplane6`. Find the equilibrium point at $(-1/3, 4/3)$. Does the data in the Equilibrium point data window agree with your findings in part (a)? *Note:* The eigenvalues of the Jacobian predict classification of the equilibrium point. In this case, the point $(-1/3, 4/3)$ is a saddle because the eigenvalues are real and opposite in sign.

c) Display the linearization. Does the equation of the linearization agree with your findings in part (a)?

19. **Linearization.** Consider the system

$$x' = x^2 + xy - 2x,$$
$$y' = 3xy + y^2 - 3y.$$

Enter the system in `pplane6`.

a) Find the equilibrium point at $(1/2, 3/2)$. Display the linearization, plot the stable and unstable orbits and several solution trajectories to create a phase portrait of the linearization.

b) Return to the PPLANE6 Display window. Select **Edit→Zoom in square**, place the mouse at the equilibrium point $(1/2, 3/2)$, then drag and release a small zoom box. On a PC, you can depress the shift key, then drag with the left mouse button to achieve the same effect. Repeat several times until you have zoomed sufficiently close to the equilibrium point. Plot the stable and unstable orbits and several trajectories to create a phase portrait of the nonlinear system near $(1/2, 3/2)$. Compare this phase portrait with the phase portrait in the linearization window.

20. **When Linearization Fails.** Most of the time, the linearization accurately predicts the behavior of a nonlinear system at an equilibrium point. There are exceptions, most notably when the the matrix A has purely imaginary eigenvalues, or when one of the eigenvalues is zero. For example, consider the system

$$x' = -y + x(x^2 + y^2),$$
$$y' = x + y(x^2 + y^2).$$

Enter the system in pplane6, set the display window so that $-1 \leq x \leq 1$ and $-1 \leq y \leq 1$, select **Arrows** for the direction field, then click **Proceed**. In the PPLANE6 Display window, select **Solutions→Show nullclines** to overlay the nullclines on the vector field and note the presence of an equilibrium point at $(0, 0)$.

a) Use Keyboard input to start a solution trajectory at $(0.5, 0)$ and note that the origin behaves as a spiral source. If the solution takes too long to stop on its own use the **Stop** button.

b) Select **Solutions→Find an equilibrium point** and find the equilibrium point at $(0, 0)$. Note that the eigenvalues of the Jacobian are purely imaginary, indicating that the linearization has a *center*, not a spiral source, at the origin. Display the linearization and draw some solution trajectories.

In Exercises 21 –30, perform steps i) and ii) for the linear system

$$\mathbf{x}' = \mathbf{Ax},$$

where $\mathbf{A}$ is the given matrix.

i) Find the type of the equilibrium point at the origin. Do this without the aid of any technology or pplane6. You may, of course, check your answer using pplane.

ii) Use pplane6 to plot several solution curves — enough to fully illustrate the behavior of the system. You will find it convenient to use the "linear system" choice from the **Gallery** menu. If the eigenvalues are real, use the "Keyboard input" option to include straight line solutions starting at ± 10 times the eigenvectors. If the equilibrium point is a saddle point, compute and plot the separatrices.

21. $\begin{bmatrix} 1 & 1 \\ -18 & 10 \end{bmatrix}$ 22. $\begin{bmatrix} 6 & -1 \\ 0 & -3 \end{bmatrix}$ 23. $\begin{bmatrix} -1 & 0 \\ 3 & -3 \end{bmatrix}$ 24. $\begin{bmatrix} 6 & 1 \\ -18 & 0 \end{bmatrix}$ 25. $\begin{bmatrix} 9 & -1 \\ 9 & 3 \end{bmatrix}$

26. $\begin{bmatrix} 9 & 1 \\ 9 & 3 \end{bmatrix}$ 27. $\begin{bmatrix} 7 & 4 \\ -10 & -5 \end{bmatrix}$ 28. $\begin{bmatrix} -1 & 0 \\ 0 & -1 \end{bmatrix}$ 29. $\begin{bmatrix} 6 & -4 \\ 18 & -6 \end{bmatrix}$ 30. $\begin{bmatrix} 2 & -2 \\ 4 & -4 \end{bmatrix}$

In Exercises 31 –34 we consider the effect of a non-linear perturbation which leaves the Jacobian at $(0, 0)$ unchanged. The theory predicts that, in (maybe very small) neighborhoods of equilibrium points where the eigenvalues of the Jacobian are non-zero and not equal, a non-linear system will act very much like the linear system associated with the Jacobian. We will use `pplane6` to verify this by performing steps i), ii), and iii) below for the system in the indicated exercise.

i) Make up a perturbation for the system that vanishes to second order at the origin. As an example, instead of the system in Exercise 21 consider the perturbed system

$$x' = x + y - xy,$$
$$y' = -18x + 10y - x^2 - y^2.$$

ii) Show that the Jacobian of the perturbed system at the origin is the same as that of the unperturbed system.

iii) Starting with the display rectangle $-5 \le x \le 5, -5 \le y \le 5$, zoom in square to smaller squares centered at $(0, 0)$ until the solution curves look like those of the linear system.

31. Exercise 21.   32. Exercise 22.   33. Exercise 23.   34. Exercise 24.

35. In contrast to Exercises 31 – 34, consider the system

$$x' = y + ax^3,$$
$$y' = -x,$$

for the three values 0, 10 and -10 of the parameter a.

a) Show that all three systems have the same Jacobian matrix at the origin. What type of equilibrium point at $(0, 0)$ is predicted by the eigenvalues of the Jacobian?

b) Use `pplane6` to find evidence that will enable you to make a conjecture as to the type of the equilibrium point at $(0, 0)$ in each of the three cases.

c) Consider the function $h(x, y) = x^2 + y^2$. In each of the three cases, restrict h to a solution curve and differentiate the result with respect to t (Recall: $dh/dt = (\partial h/\partial x)(dx/dt) + (\partial h/\partial y)(dy/dt)$). Can you use the result to verify the conjecture you made in part b)? *Hint:* Note that $h(x, y)$ measures the distance between the origin and (x, y).

d) Does the Jacobian predict the behavior of the non-linear systems in this case?

36. Consider the system

$$x' = ax + y,$$
$$y' = -x - 2y.$$

Describe what happens to the equilibrium points as a varies among the five choices -0.1, -0.01, 0, 0.01, and 0.1. Provide plots of each with enough solution curves to illustrate the behavior of the system. If the eigenvalues are real, use the **Keyboard input** option to include straight line solutions starting at ± 1 times the eigenvectors.

37. Consider the damped pendulum when the damping constant is D=0.5 versus when D=2.5.

a) Select **Gallery→pendulum** and use `pplane6` to examine the behavior of the system in each of these cases.

b) You will notice that the type of the equilibrium point at the origin is different for D=0.5 and D=2.5. Analyze the Jacobian matrix and find out at which value of D the type changes.

38. This is another example of how a system can change as a parameter is varied. Consider the system

$$x' = ax + y - x(x^2 + y^2),$$
$$y' = -x + ay - y(x^2 + y^2),$$

for $a = 0$, and $a = \pm 0.1$. Use the display rectangle $-1 \le x \le 1$, and $-1 \le y \le 1$, and plot enough solutions in each case to describe behavior of the system. Describe what happens as the parameter a varies from negative values to positive values. (This is an example of a *Hopf bifurcation*.) *Hint:* Look for something completely new, something other than an equilibrium **point**.

39. This problem involves the default system in `pplane6`, i.e., the planar system that appears when `pplane6` is first started.

   a) Find and plot all interesting features, i.e., equilibrium points, separatrices, limit cycles, and any other features that appeal to you. Make a list of such to turn in with the plot.

   b) Find the linearization of the system at $(0, 0)$. Enter that into `pplane6` and plot a few orbits.

   c) Go back to the default system. According to theory, the linearization should approximate the original system in a small enough neighborhood of $(0, 0)$. Use the **Zoom in square** option to find a small enough rectangle containing $(0, 0)$, where the behavior of the system is closely approximated by the linear system you analyzed in part b).

   d) Redo parts b) and c) for the other equilibrium points of the default system.

40. Perturbation of the default system in `pplane6` exhibits two bifurcations involving the saddle point at the origin and the equilibrium point near $(3/2, -3/2)$. To be precise we are looking at the system

$$x' = 2x - y + 3(x^2 - y^2) + 2xy,$$
$$y' = x - 3y - 3(x^2 - y^2) + Axy,$$

as the parameter A varies.

   a) Show that between $A = 2.5$ and $A = 3$, the equilibrium point changes from a sink to a source and spawns a limit cycle, i.e., a Hopf bifurcation occurs.

   b) Show that between $A = 3$ and $A = 3.2$ the limit cycle disappears. At some point in this transition, the limit cycle becomes a homoclinic orbit for the saddle point at the origin. A *homoclinic orbit* is one that originates as an unstable orbit for saddle point at $t = -\infty$, and then becomes a stable orbit for the same saddle point as $t \to +\infty$. It is extremely difficult to find the value of A for which this occurs, so do not try too hard. The process illustrated here is called a *homoclinic bifurcation*.

   c) The default system has four quadratic terms. Show that if any of the coefficients of these terms is altered up or down by as little as 0.5 from the default, a bifurcation takes place. Show that there is a Hopf bifurcation in one direction, and a homoclinic bifurcation in the other.

41. If $\mathbf{x}_0$ is a sink for the system $\mathbf{x}' = \mathbf{F}(\mathbf{x})$, then the *basin of attraction* for $\mathbf{x}_0$ is the set of points $\mathbf{y} = [y_1, y_2]^T$ which have the property that the solution curve through $\mathbf{y}$ approaches $\mathbf{x}_0$ as $t \to +\infty$. Consider the damped pendulum with damping constant $a = 0.5$. Find the basin of attraction for $[0, 0]^T$. You should indicate this region on a printed output of `pplane6`. *Hint:* It will be extremely helpful to plot the stable and unstable orbits from a couple of saddle points.

42. The system of differential equations:

$$x' = \mu x - y - x^3,$$
$$y' = x,$$

is called the *van der Pol system*. It arises in the study of non-linear semiconductor circuits, where y represents a voltage and x the current. It is in the **Gallery** menu.

   a) Find the equilibrium points for the system. Use `pplane6` only to check your computations.

   b) For various values of μ in the range $0 < \mu < 5$, find the equilibrium points, and find the type of each, i.e, is it a nodal sink, a saddle point, ...? You should find that there are at least two cases depending on the value of μ. Don't worry too much about non-generic cases. Use `pplane6` only to check your computations.

c) Use `pplane6` to illustrate the behavior of solutions to the system in each of the cases found in b). Plot enough solutions to illustrate the phenomena you discover. Be sure to start some orbits very close to $(0, 0)$, and some near the edge of the display window. Put arrows on the solution curves (by hand after you have printed them out) to indicate the direction of the motion. (The display window $(-5, 5, -5, 5)$ will allow you to see the interesting phenomena.)

d) For $\mu = 1$ plot the solutions to the system with initial conditions $x(0) = 0$, and $y(0) = 0.2$. Plot both components of the solution versus t. Describe what happens to the solution curves as $t \to \infty$.

43. *Duffing's equation* is

$$mx'' + cx' + kx + lx^3 = F(t).$$

When $k > 0$ this equation models a vibrating spring, which could be soft ($l < 0$) or hard ($l > 0$) (See Student Project #1 in Chapter 8). When $k < 0$ the equation arises as a model of the motion of a long thin elastic steel beam that has its top end embedded in a pulsating frame (the $F(t)$ term), and its lower end hanging just above two magnets which are slightly displaced from what would be the equilibrium position. We will be looking at the unforced case (i.e. $F(t) = 0$), with $m = 1$. The system corresponding to Duffing's equation is available in the **Gallery** menu.

a) This is the case of a hard spring with $k = 16$, and $l = 4$. Use `pplane6` to plot the phase planes of some solutions with the damping constant $c = 0, 1,$ and 4. In particular, find all equilibrium points.

b) Do the same for the soft spring with $k = 16$ and $l = -4$. Now there will be a pair of saddle points. Find them and plot the stable/unstable orbits.

c) Now consider the case when $k = -1$, and $l = 1$. For each of the cases $c = 0$, $c = 0.2$, and $c = 1$, use `pplane6` to analyze the system. In particular, find all equilibrium points and determine their types. Plot stable/unstable orbits where appropriate, and plot typical orbits.

d) With $c = 0.2$ in part c), there are two sinks. Determine the basins of attraction of each of these sinks. Indicate these regions on a print out of the phase plane.

44. A wide variety of phenomena can occur when an equilibrium point is completely degenerate, i.e., when the Jacobian is the zero matrix. We will look at just one. Consider the system

$$x' = xy,$$
$$y' = x^2 - y^2.$$

a) Show that the Jacobian at the origin is the zero matrix.

b) Plot the solutions through the six points $(0, \pm 1)$, and $(\pm\sqrt{2}, \pm 1)$. Plot additional solutions of your choice.

c) Compare what you see with the behavior of solutions near a saddle point.

Appendix: A Summary of Equilibrium Point Messages

Pplane6 will recognize the types of many equilibrium points, but it is very conservative in its approach, and will often admit that it is unsure. Since `pplane6` examines an equilibrium point numerically, and because numerical algorithms always make small truncation errors, such a conservative approach is required. In particular, `pplane6` will only report information that is generic, i.e. information that will not change if the data on which it is based is changed by a small amount. Here is a complete list of the messages that can appear describing the equilibrium point, followed by an indication of what the message means.

```
There is a saddle point at ...
```

The eigenvalues are real and have different signs.

There is a nodal sink at ...

The eigenvalues are real, distinct, and negative. In some books this is called an *improper node*, a *stable improper node*, or a *stable node*.

There is a nodal source at ...

The eigenvalues are real, distinct, and positive. In some books this is called an *improper node*, an *unstable improper node*, or an *unstable node*.

There is a spiral sink at ...

The eigenvalues are complex conjugate and the real parts are negative. This is sometimes called a *spiral point*, a *stable spiral point*, or a *stable focus*.

There is a spiral source at ...

The eigenvalues are complex conjugate and the real parts are positive. This is sometimes called a *spiral point*, an *unstable spiral point*, or an *unstable focus*.

There is a spiral equilibrium point at ...

Its specific type has not been determined.

The eigenvalues are complex conjugate, with non-zero imaginary part. However, the real part is too close to 0 to decide whether this is a source, a sink, or a center. In particular, any true center (i.e. conjugate, purely imaginary eigenvalues) will be included in this category. A center is also called a *focus* in some books.

There is a source at ...

Its specific type has not been determined.

The eigenvalues have positive real part, but the eigenvalues are too close together to decide the exact type of the source. This could be a spiral source, a nodal source, or a degenerate source for which the Jacobian has equal eigenvalues. An equilibrium point of this type is unstable, but not much more can be said.

There is a sink at ...

Its specific type has not been determined.

The eigenvalues have negative real part, but the the eigenvalues are too close together to decide the exact type of the sink. This could be a spiral sink, a nodal sink, or a degenerate sink for which the Jacobian has equal eigenvalues. An equilibrium point of this type is asymptotically stable, but not much more can be said.

There is an equilibrium point at ...

Its specific type has not been determined

At least one of the eigenvalues is so close to 0 that it is not possible to decide anything about the type of the equilibrium point. Closer analysis is required to describe the behaviour of the solutions.

It can be seen that `pplane6` will label an equilibrium point only when it is sure. Of course, more precise information about an equilibrium point can sometimes be garnered by looking at the Jacobian matrix. The Jacobian as provided by `pplane6` is a numerical approximation to the actual matrix of derivatives. Numerical approximations to derivatives are frequently inaccurate, so, especially in cases were there is doubt, the actual Jacobian should be computed and analyzed (either by hand or with the Symbolic Toolbox). The power of computers is still not sufficient to eliminate the need to do algebra.

However, looking at the Jacobian may not suffice, since this amounts to looking at the linearization of the system, and not the actual system. There are systems for which the linearization does not tell the whole story. Consequently, in the cases when `pplane6` cannot decide the precise type, the ultimate test of the type is what the solution curves actually do. Even then it is entirely possible that visualization will not provide the answer, and a rigorous analysis will be necessary.

Index

Items that appear in typewriter font, such as dfield, refer to MATLAB commands or variables.

overdetermined, 170.

underdetermined, 170.

systems of first order equations, 108.

systems of linear equations, 154.

systems of linear equations, solving, 157.

systems of odes, 93.

systems with only one eigenvalue, 198.

systems

autonomous, 212.

complex solutions, 193.

constant coefficients 181, 187.

homogeneous, 181, 187.

linear, 181, 217.

nonlinear, 219, 231.

planar, 212.

solving symbolically, 146.

the general case, 200.

vector form, 108.

with repeated eigenvalues, 197.

tan, 3.

tangent line, 219.

tanh, 4.

Taylor's series, 219.

teddy bears, 102.

T$_{E}$X notation, 39, 79.

in odesolve, 132.

text objects in dfield6, 42.

text strings in MATLAB, 22.

title, 21, 147, 151.

Toolbar, 8.

total response, 18.

trace, 230.

trace-determinant plane, 230.

transient, 128.

transient behavior, 116.

transient response, 18.

transpose, 13.

transpose, conjugate, 13.

troubleshooting function M-files, 55.

truncation errors, 235.

tspan, 62.

type, 56.

type of equilibrium point, 235.

underdetermined systems of equations, 170.

uniqueness, 33, 46.

unknown vector, 155.

unstable, 37, 45, 130.

unstable focus, 236.

unstable improper node, 236.

unstable node, 236.

unstable solution curve, 224.

unstable spiral point, 236.

using clipboards with dfield6, 42.

van der Pol system, 124, 226, 234.

variables, scaling, 97.

vector, 12.

vector field, 40, 94, 212.

vector form of systems, 108.

vector of unknowns, 155.

vector, scalar multiple of, 13.

vectors, appending, 66.

vibrating spring, 126, 226, 235.

Volterra, Umberto, 127.

which, 56.

whos, 110, 139, 140.

Window settings, 40.

window, DFIELD6 Display, 29–43, 85–91, 100.

window, DFIELD6 Setup, 28–43, 79–95.

window, PPLANE6 Display, 95–100, 213–232.

window, PPLANE6 Setup, 95–104, 213–224.

workspace, function, 114.

xlabel, 21, 147, 151.

xtick, 83.

ygrid, 121.

ylabel, 21, 147, 151.

ytick, 121.

zoom back, 35, 96.

zoom box, 34.

zoom in, 32, 96, 223.

zoom in square, 223, 232, 233, 234.